CAMBRIDGE MONOGRAPHS ON PHYSICS

GENERAL EDITORS

B. H. FLOWERS, F.R.S.
Langworthy Professor of Physics in the University of Manchester

J. M. ZIMAN, D.PHIL.
Fellow of King's College, Cambridge

SUPERCONDUCTIVITY

T0296132

H. KAMERLINGH ONNES, 1853–1926

[*From a drawing by his nephew, H. Kamerlingh Onnes*]

SUPERCONDUCTIVITY

BY

D. SHOENBERG, F.R.S.

*Fellow of Gonville & Caius College,
Cambridge*

CAMBRIDGE
AT THE UNIVERSITY PRESS
1965

CAMBRIDGE UNIVERSITY PRESS
Cambridge, New York, Melbourne, Madrid, Cape Town,
Singapore, São Paulo, Delhi, Tokyo, Mexico City

Cambridge University Press
The Edinburgh Building, Cambridge CB2 8RU, UK

Published in the United States of America by
Cambridge University Press, New York

www.cambridge.org
Information on this title: www.cambridge.org/9780521092548

First edition (in Cambridge Physical Tracts) 1938
Second edition (in Cambridge Monographs on Physics) 1952
Reprinted with additional appendix 1960
Reprinted 1962, 1965
Re-issued 2011

A catalogue record for this publication is available from the British Library

ISBN 978-0-521-09254-8 Paperback

PREFACE TO SECOND EDITION

Since publication of the first edition of this monograph there have been considerable advances both in our knowledge of the facts of superconductivity and in our understanding of them. On the experimental side, the slight penetration of a magnetic field into a superconductor and its concomitant effects, such as high-frequency resistance, have been studied in a variety of ways, direct evidence has been provided for theories of the structure of the intermediate state, and much useful data has been accumulated about the thermodynamic properties of individual superconductors and about the existence of superconductivity in elements, alloys and compounds. On the theoretical side, the new experimental results have stimulated interest in the already existing phenomenological theories of F. and H. London and of Gorter and Casimir, and have led to development of certain aspects of these theories to the point where, in spite of their value in guiding research, they begin to show themselves inadequate in describing the facts. At the same time there have been several new attempts at fundamental theories, which make it probable that superconductivity will not for much longer remain in its hitherto impregnable position of an unsolved mystery of science.

All these advances have necessitated a complete revision of the first edition of the monograph. Much of the discussion of the magnetic and thermodynamic properties of macroscopic superconductors (most of Chapters I–VI of the first edition) still remains valid, and though now presented in a different order has not required very substantial modification except for the addition of new sections on thermal conductivity and thermoelectric effects. The material of Chapter VII dealing with penetration effects and of part of Chapter III dealing with the structure of the intermediate state has, however, not merely been superseded but has been replaced by a much larger body of material, and now occupies nearly half of the present monograph (Chapters IV and V and parts of Chapter VI). In order to discuss this new material adequately an account of the phenomenological theories has been added (Chapter VI), and for completeness a brief and entirely descriptive account of some of the recent attempts at a fundamental theory has also been added.

In keeping with the slight change of emphasis accompanying the change from 'Tracts' to 'Monographs', rather more details of experimental facts—in the forms of graphs and tables—have been included, and the Appendix on numerical data somewhat enlarged. It is hoped that in its new form the monograph may be found useful not only as a general account of the subject, but also as a reference manual for the research worker.

I owe a great debt of gratitude to my colleague Dr A. B. Pippard for introducing me to many new ideas on which I have drawn freely in writing this monograph and for much helpful criticism and stimulating discussion. I should like also to thank Dr R. G. Chambers for his help in correcting the proofs and for many valuable suggestions, Mr T. E. Faber, Mr C. G. Kuper, Dr J. M. Lock and Mr G. T. Pullan for reading various parts of the monograph in manuscript and in proof, and all the above-mentioned and Dr J. R. Clement, Dr B. B. Goodman, Dr J. K. Hulm, Dr K. Mendelssohn, Dr E. Mendoza, Dr J. L. Olsen and Dr A. Wexler for their kindness in letting me have details of their work in advance of publication and for allowing me to quote from their communications. I have also to thank the Royal Society for permission to reproduce a number of diagrams published in the Proceedings.

D. SHOENBERG

ROYAL SOCIETY MOND LABORATORY
CAMBRIDGE
12 *July* 1951

NOTE

In this reprinting the opportunity has been taken of adding a brief note, with bibliography, on developments since 1952 (see p. 240).

D. SHOENBERG

CAMBRIDGE
June 1959

PREFACE TO FIRST EDITION

During the last few years, particularly since 1933, research on superconductivity has taken rather a new turn; previously it had always been tacitly assumed that the only essential feature of the superconducting state was infinite conductivity, and that the magnetic behaviour was a secondary feature which could be predicted from this property alone. The now classic experiment of Meissner and Ochsenfeld showed, however, that the predictions were sometimes quite wrong, and this discovery, together with the results of the recent work arising out of it, has led to a considerable revision of the phenomenological aspect of superconductivity, and has made more intelligible the fact that this change is not appreciably reflected except in the electrical and magnetic properties of the metal.

At the time when Kamerlingh Onnes first discovered superconductivity—1911—no proper theory of metals existed, and it was not surprising that this new phenomenon also could not be explained; since then, however, although wave-mechanics has found a satisfactory qualitative explanation for most metallic phenomena, superconductivity has remained as anomalous as ever from the theoretical point of view. The recent developments have not materially changed this position, but their importance is that they have at least made possible a coherent statement of what it is that the theory has to explain. This statement we have attempted to set out in the present survey.

Although the survey has been written mainly from the point of view of the recent developments, showing how it has become possible to correlate some of the properties of superconductors and thus to distinguish between fundamental and secondary features, it is intended also as a fairly comprehensive summary of superconductive phenomena in general. In describing the various experimental results, our aim has been to make clear the essential principles involved, rather than the details of the methods by which the results have been obtained; similarly, to avoid confusing the main issues, no numerical data have been introduced into the text except for illustrative purposes, but these data are collected instead in an appendix for reference.

I wish to thank Prof. Landau for introducing me to many new ideas about the intermediate state (some as yet unpublished), to the idea that the zero resistance of a superconductor is a consequence of an exactly zero induction, and to the simple thermodynamical treatment of Chapter VI, and also for much detailed criticism which exposed (and removed) a number of theoretical prejudices originally held. I should like also to thank Dr Alexeevski, Dr Mendelssohn, Dr Misener, Dr Pomerantchuk and Prof. Shalnikov for their kindness in communicating details of their various researches in advance of publication, and allowing me to quote from their communications.

D. SHOENBERG

INSTITUTE FOR PHYSICAL PROBLEMS,
ACADEMY OF SCIENCES OF THE U.S.S.R., MOSCOW
February, 1938

NOTE ON THE 1965 IMPRESSION

Some excuse is perhaps needed for a new reprint of a monograph on superconductivity written before the almost explosive burst of new activity which followed the development of a fundamental theory and has been stimulated by the growing interest in possible technological applications of superconductors. The phenomenological description of the magnetic and thermodynamic properties and of size effects in superconductors, which form the main theme of the monograph, tend nowadays to be taken rather for granted, though they are topics which contain many pitfalls for the unwary and yet must be properly understood before the new ideas can be really appreciated. This monograph may then still serve as a useful introduction for the newcomer to the field.

The state of the whole subject was comprehensively reviewed at the Colgate conference in August 1963 and the conference report (*Rev. Mod. Phys.* **36**, 1, 1964) contains detailed references to recent literature. A brief bibliography of important review articles up to 1959 will be found in Appendix IV (p. 240).

D. SHOENBERG

CAMBRIDGE
8 June 1964

PREFACE TO FIRST EDITION

During the last few years, particularly since 1933, research on superconductivity has taken rather a new turn; previously it had always been tacitly assumed that the only essential feature of the superconducting state was infinite conductivity, and that the magnetic behaviour was a secondary feature which could be predicted from this property alone. The now classic experiment of Meissner and Ochsenfeld showed, however, that the predictions were sometimes quite wrong, and this discovery, together with the results of the recent work arising out of it, has led to a considerable revision of the phenomenological aspect of superconductivity, and has made more intelligible the fact that this change is not appreciably reflected except in the electrical and magnetic properties of the metal.

At the time when Kamerlingh Onnes first discovered superconductivity—1911—no proper theory of metals existed, and it was not surprising that this new phenomenon also could not be explained; since then, however, although wave-mechanics has found a satisfactory qualitative explanation for most metallic phenomena, superconductivity has remained as anomalous as ever from the theoretical point of view. The recent developments have not materially changed this position, but their importance is that they have at least made possible a coherent statement of what it is that the theory has to explain. This statement we have attempted to set out in the present survey.

Although the survey has been written mainly from the point of view of the recent developments, showing how it has become possible to correlate some of the properties of superconductors and thus to distinguish between fundamental and secondary features, it is intended also as a fairly comprehensive summary of superconductive phenomena in general. In describing the various experimental results, our aim has been to make clear the essential principles involved, rather than the details of the methods by which the results have been obtained; similarly, to avoid confusing the main issues, no numerical data have been introduced into the text except for illustrative purposes, but these data are collected instead in an appendix for reference.

I wish to thank Prof. Landau for introducing me to many new ideas about the intermediate state (some as yet unpublished), to the idea that the zero resistance of a superconductor is a consequence of an exactly zero induction, and to the simple thermodynamical treatment of Chapter VI, and also for much detailed criticism which exposed (and removed) a number of theoretical prejudices originally held. I should like also to thank Dr Alexeevski, Dr Mendelssohn, Dr Misener, Dr Pomerantchuk and Prof. Shalnikov for their kindness in communicating details of their various researches in advance of publication, and allowing me to quote from their communications.

D. SHOENBERG

INSTITUTE FOR PHYSICAL PROBLEMS,
ACADEMY OF SCIENCES OF THE U.S.S.R., MOSCOW
February, 1938

NOTE ON THE 1965 IMPRESSION

Some excuse is perhaps needed for a new reprint of a monograph on superconductivity written before the almost explosive burst of new activity which followed the development of a fundamental theory and has been stimulated by the growing interest in possible technological applications of superconductors. The phenomenological description of the magnetic and thermodynamic properties and of size effects in superconductors, which form the main theme of the monograph, tend nowadays to be taken rather for granted, though they are topics which contain many pitfalls for the unwary and yet must be properly understood before the new ideas can be really appreciated. This monograph may then still serve as a useful introduction for the newcomer to the field.

The state of the whole subject was comprehensively reviewed at the Colgate conference in August 1963 and the conference report (*Rev. Mod. Phys.* **36**, 1, 1964) contains detailed references to recent literature. A brief bibliography of important review articles up to 1959 will be found in Appendix IV (p. 240).

D. SHOENBERG

CAMBRIDGE
8 June 1964

CONTENTS

Frontispiece: H. Kamerlingh Onnes (1853–1926)

CHAPTER I
Introduction

CHAPTER II
Magnetic properties of macroscopic superconductors

CHAPTER III
Thermodynamic and other thermal properties

CHAPTER IV
Structure of the intermediate state

CHAPTER V

The depth of penetration of a magnetic field
into a superconductor

CHAPTER VI

Theoretical aspects

INTRODUCTION

1.1. Discovery of superconductivity

It was known a long time ago that the resistance of a metal decreases with temperature, and soon after Kamerlingh Onnes in 1908 had liquefied helium for the first time he decided to investigate how the resistance varied in the newly available range of low temperatures below 4·2° K. The theory of electrical resistance was then in rather a rudimentary state, and any of the three modes of

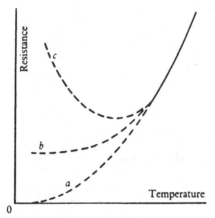

Fig. 1. Possible forms of temperature variation of resistance (schematic).

behaviour sketched in fig. 1 seemed possible. Thus *a* would occur if the resistance was due entirely to obstruction of electronic paths by thermal vibrations, *b* would occur if obstruction by impurities and imperfections was important, while *c* would occur if the number of free electrons available to carry the current fell off rapidly at low temperatures due to 'condensation' on the atoms.

The new experimental results (Onnes, 1911 *a*, *b*, *c*) (see fig. 2) were presented by Onnes in an interesting paper* (Onnes, 1913 *a*)

* This paper gives a general survey of the new results, which were only briefly and provisionally reported in the earlier papers. The superconductivity of mercury was suggested by the experiments reported in April 1911 (Onnes, 1911 *b*) and established more definitely in May 1911 (Onnes, 1911 *c*).

to the Third International Congress of Refrigeration held in September 1913 at Chicago: 'I already inclined to the idea that had been expressed by Dewar, that resistance would tend to vanish at the absolute zero itself, when the experiments with liquid helium brought quite a revelation. The resistance of very pure platinum became constant instead of passing through a minimum or of tending to vanish at the absolute zero.' The constant value of the resistance turned out to depend strongly on the impurity content of the metal, and indeed such an impurity effect could have been anticipated, for nearly 50 years earlier, in 1864,

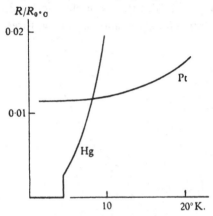

Fig. 2. Variation of $R/R_{0°C.}$ (ratio of the resistance to resistance at 0° C.) of platinum and mercury (Onnes, 1913 *a*).

Matthiessen had found that specimens of the same metal differing in purity differed in resistance by amounts which did not vary with temperature.

Onnes then explains that since it seemed as if it was only impurities which prevented the resistance of platinum and gold disappearing, even 'using the purest gold of any mint in the world', he decided to experiment with the 'only metal which one could hope to get into wires of a higher state of purity, viz. mercury....It could be foretold that the resistance of a wire of solid mercury would be measurable at the boiling point of helium but would fall to inappreciable values at the lowest temperatures which I could reach. With this beautiful prospect before me there was no more question of reckoning with difficulties. They were overcome and the result of the experiments was as convincing as could be hoped. No doubt was left of the existence of a new state of mercury in which its resistance has practically vanished....Mercury has passed into a new state, which on account of its extraordinary electrical properties may be called the superconductive state.'

It is curious that the discovery of superconductivity should have come about in this way, for Onnes himself soon showed that the confirmation of his predictions about the behaviour of mercury was only apparent. Thus he found that considerable impurities added to mercury did not in fact inhibit the drop to zero resistance, i.e. that the zero resistance was not just a matter of using a very pure metal. At the same time more careful measurements (Onnes, 1911 d) showed that the resistance fell to zero much more sharply than the fall Onnes had foretold (which was according to a formula based on what is now known to be an unsound theory).

The discovery of superconductivity opened up a whole series of problems about the scope and nature of the new phenomenon. As regards the scope of superconductivity, it has been found that twenty-one metallic elements and a large number of alloys become superconducting, the transition temperature being characteristic of the particular metal, and varying from as low as 0·35° K. for hafnium up to about 8° K. for niobium (some alloys have higher transition temperatures, the highest known being about 15·5° K. for niobium nitride). The known superconducting elements fall roughly into two groups in the periodic system (see Appendix I, Table VIII), which suggests that superconductivity is probably not a universal property (see also p. 38); but since every new advance in the lowering of temperature* has revealed new superconductors, we cannot yet be certain that superconductivity is, indeed, limited only to certain metals. Since it is impossible to reach the absolute zero, this question can be definitely settled only by the development of a theory of the origin of superconductivity; unless, of course, further advances in lowering temperature should show that *all* metals do, indeed, become superconducting.

Attempts have been made to associate superconductivity with some other property of the metal, but apart from the fact that the

---

* For instance, application of the adiabatic demagnetization technique (Kürti and Simon, 1935; Goodman, 1951 a).

superconducting elements all have about the same atomic volume (Clusius, 1932), and perhaps the rule suggested above, that super-conductivity is confined to certain groups of the periodic system, most suggested empirical rules of this kind have broken down with the discovery of new superconductors.* For instance, it was thought that only 'soft' metals with low melting-points became superconducting, until it was found that the typical 'hard' metals tantalum and niobium also became superconducting, and, more-over, with relatively high transition temperatures. Similarly, no particular type of crystal lattice seems to be necessary for super-conductivity, since nearly all the types are represented in the list of known superconductors. By this we do not, of course, mean to imply that the crystal lattice is of less importance than the atomic species of the metal, for just as with ordinary conductivity, any change of crystal lattice for a metal of given atomic species can have a profound effect. This is strikingly illustrated by the case of tin; thus white tin is a typical superconductor, while grey tin, which differs from it only in its crystal lattice, does not become superconducting down to $1.3°$ K., the lowest temperature tried (Sharvin, 1945).

As already mentioned, an important characteristic of the loss of resistance in the superconducting transition is the sharpness of the transition. Careful experiments by de Haas and Voogd (1931c) suggested that in 'ideal' conditions the transition would be practically discontinuous, the observed breadth being due to slight inhomogeneity of purity or strain or to the effect of a finite measuring current (see § 4.7); some of their results are shown in fig. 3. For very small measuring currents they found that the breadth of transition of tin could be reduced to as little as $0.001°$ K. in single crystals of the highest purity. Pippard (1950c) has suggested that even in a perfect specimen it may not be possible to reduce the breadth of transition much further, owing to a domain structure of a fundamental kind (see § 6.3.5).

* Some recent theories of superconductivity (see § 6.4) have suggested specific criteria which are fairly successful. For instance, in Fröhlich's theory (Fröhlich, 1950) a certain dimensionless combination of room temperature conductivity, degeneracy parameter, velocity of sound and atomic mass has to exceed unity for the metal to be a superconductor. The fact that this rule fails for several metals may be regarded as due to the over-simplifications of the theory in its present form.

The use of the name 'superconductivity' by Kamerlingh Onnes
has become more and more justified with the passage of time. In
his early experiments Onnes (1913 b), using a potentiometer

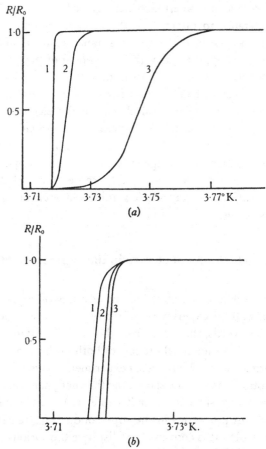

Fig. 3. Influence of specimen quality on sharpness of superconducting transition (de Haas and Voogd (1931 c), but temperature scale corrected to 1949; R_0
is the resistance at $4 \cdot 2°$ K.). (a) Influence of specimen quality: (1) pure tin
single crystal; (2) pure tin polycrystal; (3) less pure tin polycrystal. (b) Influence
of measuring current for pure tin single crystal of diameter $0 \cdot 32$ mm.: (1) 40 mA.;
(2) 20 mA.; (3) 10 mA.

method, was already able to show by passing high currents through
the wire that the resistance had fallen by a factor of at least 10^6
in the transition. Since then by means of the ingenious method
(also due to Onnes) of looking for a decay of the 'persistent'

current in a ring (see § 2.6), it has been possible to show that the resistance in the superconducting state has less than 10^{-12} of its value immediately above the transition (Onnes, 1914b; Grassmann, 1936). In fact, no experiment has succeeded in revealing any trace of resistance to direct current in the superconducting state, and it can be assumed that the resistance really is zero.* This is equivalent to supposing that the component of the electric field E in the direction of current flow is zero.

Presumably the complete and abrupt loss of resistance must be a consequence of some fundamental change in the electronic or atomic structure of the metal, and the experiments to elucidate the nature of this change can be divided roughly into two classes: (a) those to test whether this change is reflected in any properties of the metal other than its resistance, and (b) those to test whether the change can be affected in any way (e.g. inhibited or encouraged) by various physical agents.

1.2. Properties which do not change in the superconducting transition

For a long time one of the most baffling features of superconductivity was that all the experiments of the type (a) gave negative results, or in other words, that apart from the loss of resistance the metal appeared to have identical properties both in the superconducting and normal states. More recent experiments have revealed important exceptions, and at the same time thermodynamical considerations have, as we shall see later (Chapter III), made the lack of change of certain properties in the superconducting transition more understandable. It is convenient to list here the various properties which have so far been investigated with negative results.

1.2.1. The X-ray diffraction pattern is the same both above and below the transition temperature, which shows that no change of the crystal lattice is involved. The absence of any appreciable change in the intensity distribution also shows that the change in the electronic structure must be very slight (Keesom and Onnes, 1924).

---

* Holm and Meissner (1932) have shown also that there is no resistance at a clean contact between two superconductors.

1.2.2. There is no appreciable change in the reflectivity of the metal either in the visible or the infra-red region, although the optical properties are usually closely connected with the electrical conductivity (Hirschlaff, 1937; Daunt, Keeley and Mendelssohn, 1937; Ramanathan, 1952).

1.2.3. There is no change in the absorption of fast or slow electrons (McLennan, McLeod and Wilhelm, 1929; Meissner and Steiner, 1932), and the photoelectric properties are also unchanged (McLennan, Hunter and McLeod, 1930).

1.2.4. In the absence of a magnetic field there is no latent heat and no change of volume at the transition. We shall see in Chapter III that this is connected thermodynamically with the magnetic properties.

1.2.5. The elastic properties (de Haas and Kinoshita, 1927) and the thermal expansion (McLennan, Allen and Wilhelm, 1931 a) do not seem to change in the transition, but as mentioned in § 1.3.2, and discussed more fully in § 3.5, this is probably due to inadequate sensitivity of the experimental methods.

1.3. Properties which do change in the superconducting transition

The important exceptions mentioned above, i.e. the properties which do change either abruptly or gradually when the metal becomes superconducting, will be discussed in detail in the following chapters, but it will be useful to mention them here in brief outline without comment.

1.3.1. The magnetic properties (Chapter II) undergo a change no less remarkable than that of the electrical properties. In the pure superconducting state practically no magnetic flux is able to penetrate the metal, which thus behaves as if it had zero permeability or a strong diamagnetic susceptibility. Because of this, specimen shape plays an important role, and when superconductivity is destroyed by a magnetic field (see § 1.4.1) the magnetic behaviour becomes complicated for any shape except that of the long cylinder parallel to the field. In such circumstances the specimen breaks up into a mixture of superconducting and normal regions known as the intermediate state (see § 2.5 and Chapter IV).

1.3.2. The specific heat changes discontinuously at the transition temperature, and in the presence of a magnetic field there is also a latent heat of transition. There is also a small change of volume when the transition occurs in a magnetic field. All these features find a detailed thermodynamic explanation (see Chapter III) in terms of the magnetic properties. This thermodynamic theory predicts also discontinuities in the thermal expansion and elastic properties, but, as mentioned above, these have not yet been observed owing to the smallness of the changes.

1.3.3. All the thermoelectric effects disappear in the super-conducting state (see § 3.7).

1.3.4. The thermal conductivity (see § 3.6) changes discontinuously when superconductivity is destroyed in a magnetic field, though its order of magnitude remains the same. It is lower in the superconducting state for a pure metal, but higher for certain alloys. In the absence of a magnetic field there is no discontinuity.

1.4. Influences of external agents on superconductivity

We now turn to the results of experiments of type (b) (p. 6), and again list the most important results, postponing where necessary a full discussion to later chapters.

1.4.1. *Magnetic field.* If a magnetic field is applied parallel to the length of a long superconducting wire, the resistance of the wire is suddenly restored at a definite field strength which depends on the temperature, and is characteristic of the particular metal concerned; this field is known as the 'critical field'.* The restoration of resistance is, however, abrupt only if the metal is perfectly pure and free from strains and if the current used for measuring the resistance is vanishingly small. The absence of impurities and strains is important because these slightly change the critical field, and consequently different regions of the specimen may have different critical fields if impurities or strains are present, so that the transition is blurred. The influence of the measuring current, and the necessity for the particular geometrical conditions specified

---

* The discovery of the restoration of resistance by a magnetic field was made by Onnes (1914a); the detailed character of the phenomenon, however, became clear only gradually (see, for instance, Tuyn and Onnes (1926), de Haas and Voogd (1931a)).

above, will be explained later. We may say, then, that a lack of sharpness in the restoration of resistance by a magnetic field at a given temperature (or, what is equivalent, by an increase of temperature at a given field—which may in particular be zero) is in general a secondary feature, and in our treatment we shall assume that the conditions are ideal, and that there is a perfectly definite critical field at a given temperature.*

The relation between the critical field H_c and the temperature is of great importance for characterizing the properties of any particular superconductor. It may often be represented with fair accuracy by a parabola with the equation

$$H_c = H_0(1 - (T/T_c)^2).\qquad(1.1)$$

This parabolic relation is not in general exactly true, but it provides a convenient representation which is sufficiently accurate for many purposes. The detailed data for the individual superconductors will be found in Appendix I (fig. 77 and Table IX).

For the 'soft' group of superconductors the initial slopes of the H_c-T curves at $T = T_c$ do not differ greatly from each other and are very roughly of order 100 gauss per degree (i.e. $H_0 \sim 50T_c$). For 'hard' superconductors the initial values of dH_c/dT appear to be appreciably higher, but it is not certain whether the distinction is fundamental; for instance, experiments on thorium (Shoenberg, 1940b) have shown that the H_c-T curve depends strongly on purity, and a good specimen gives a value of dH_c/dT not very different from that of a typical soft superconductor.†

We shall see in Chapter III that the H_c-T curve has a thermo-dynamic significance very similar to that of a p-T diagram for an ordinary phase transition such as melting or boiling. The finite slope of the H_c-T curve at $T = T_c$ is thermodynamically connected with the absence of a latent heat of transition, while the fact that

---

* MacDonald and Mendelssohn (1949) find that in certain circumstances the resistance of lead, tin and mercury wires is restored only over an appreciable field interval, but since their results for tin are difficult to reconcile with other data, which indicate sharp transitions, we shall suppose that the spread-out transitions arise from secondary causes. The possibility that the ideal transition is blurred must, however, not be overlooked.

† Daunt (1950) suggests that there may be a relation of the form $H_0 = CT_c^{\frac{3}{2}}$, where C is constant as between superconductors in the same group but has different values for the soft and hard groups. Recent data of Goodman and Mendoza (1951) do not, however, lend support to this suggestion.

it becomes flat as absolute zero is approached proves to be in accordance with Nernst's theorem.

1.4.2. *Current strength.* An interesting consequence of the existence of a critical magnetic field is that there is also a critical strength for the current flowing in a superconductor. The disturbance of superconductivity by a current was actually discovered (Onnes, 1913 a, c) before that by a magnetic field; a striking manifestation of the effect was the melting of a lead wire immersed in liquid helium. When the critical current was exceeded the Joule heat could not be removed fast enough owing to the formation of gas bubbles, and the temperature rose rapidly to the melting-point. At first Onnes attributed this restoration of resistance to so-called 'bad places' in the wire, but after the discovery of a critical magnetic field, Silsbee (1916) pointed out that the effect of a current in restoring the resistance might be merely due to the magnetic field which it produced. This hypothesis has since been verified in detail (see, for instance, Tuyn and Onnes (1926), Scott (1948)), thus showing that the current effect is merely a secondary feature. Some other aspects of the destruction of superconductivity by a current will be discussed in § 4.7.

1.4.3. *Stress.* The transition temperature can be changed by a stress, and usually (though not always) a stress which increases the dimensions increases the transition temperature; the order of magnitude of dT_c/dp is 10^{-10}° K. dyne^{-1} cm.^{2}, so that sensitive methods are required to detect the effect. Associated with this effect there is also a slight influence of stress on the critical magnetic field at a given temperature. These phenomena are discussed in more detail in § 3.5, where it is shown how they are thermodynamically connected with changes of volume and of thermal expansion and elastic constants in the superconducting transition.

1.4.4. *Impurity.* Addition of chemical impurities or the introduction of plastic deformation modifies nearly all the superconducting properties, and particularly the magnetic properties, in a complicated way (see § 2.8).

1.4.5. *Size.* Reduction of the size of the specimen below about 10^{-4} cm. modifies the superconducting properties in many important respects. The magnetic permeability of a small specimen is no longer zero, and varies with temperature (increasing as the temperature approaches T_c), while the critical magnetic field becomes greater than that of the bulk material. These effects, of which a full account is given in Chapter V, are associated with a temperature-dependent penetration depth of order 10^{-6} to 10^{-5} cm. of a magnetic field into a superconductor.

1.4.6. Frequency. The zero resistance of a superconductor is modified at very high frequencies of alternating current. Up to 10^7 cyc./sec. the resistance is still zero within experimental accuracy, but for 10^9 cyc./sec. H. London (1940) showed that there was a considerable resistance even below T_c. A detailed investigation

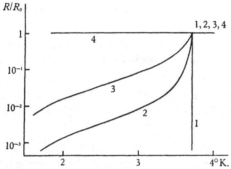

Fig. 4. Temperature variation of resistivity of tin at various frequencies (partly based on Pippard (1950a), partly schematic). (1) d.c. (2) $1\cdot2 \times 10^9$ cyc./sec. (3) $9\cdot4 \times 10^9$ cyc./sec. (4) 2×10^{13} cyc./sec.

has been made by Pippard (1947a, b, c, 1950a, b) at $1\cdot2 \times 10^9$ and $9\cdot4 \times 10^9$ cyc./sec., and some of his results are shown in fig. 4. It will be seen that the transition temperature is unaffected by the frequency, and that at $0°$ K. the resistance approaches a value which is less than 10^{-3} of R_0, the resistance in the normal state. Some results of Maxwell, Marcus and Slater (1949) suggest that a larger residual resistance remains at $0°$ K. for $2\cdot4 \times 10^{10}$ cyc./sec., but it is possible that this may be a secondary effect due to imperfect surface conditions.* At infra-red frequencies (of order 2×10^{13} cyc./sec.) the absorption of the metal (i.e. its resistivity) is the same and independent of temperature both in the normal and superconducting states (see § 1.2.2 and fig. 4), so it is probable that somewhere in the

* Other recent investigations at high frequencies have been made by Fairbank (1949), Simon (1949, 1950), Khaikin (1950) and Grebenkemper and Hagen (1951).

awkward range between infra-red and centimetre waves an absorption mechanism at absolute zero must set in. The nature of the absorption mechanism which is responsible for the resistivity at finite temperatures at high frequencies found by London and Pippard will be discussed in § 6.3.3, and some other aspects of the high-frequency properties in §§ 5.3.3, 5.3.4, 5.4 and 5.5.

1.4.7. *Isotopic constitution.* The effect of isotopic constitution was first studied by Onnes and Tuyn (1922*b*), who found that ordinary lead (atomic weight 207·30) and uranium lead (atomic weight 206·06) became superconducting at transition temperatures which did not differ by more than 0·025° K. Quite recently a positive effect of isotopic constitution on transition temperature has been found in mercury* and tin,† and it has been established that the transition temperature of a pure isotope varies to a good approximation as the square root of its atomic mass; the negative finding of Onnes and Tuyn is consistent with this result, since only 0·02° K. change could have been expected. As we shall see in § 6.4 this isotopic effect is predicted by Fröhlich's and Bardeen's theories of superconductivity.

Careful measurements of the H_c-T curves of the tin isotopes (Lock, Pippard and Shoenberg, 1951) show that in going from one isotope to another, the shape of the curve is unaltered, but the scales of ordinates and abscissae are changed by equal factors. Thus, for instance, the value of H_0, the critical field at 0° K., varies with isotopic mass in exactly the same way as does T_c. From the thermodynamic considerations of Chapter III, it can be deduced from this 'similarity' property that the electronic specific heat of the *normal* metal is independent of isotopic mass; this is just what is to be expected from the theory of metals.

* Maxwell (1950*a*), Reynolds, Serin, Wright and Nesbitt (1950), Reynolds, Serin and Nesbitt (1951).
† Maxwell (1950*b*), Allen, Dawton, Lock, Pippard and Shoenberg (1950), Lock, Pippard and Shoenberg (1951), Allen, Dawton, Bär, Mendelssohn and Olsen (1950).

CHAPTER II

MAGNETIC PROPERTIES OF MACROSCOPIC SUPERCONDUCTORS

2.1. Introduction

For many years after the discovery of superconductivity it was tacitly assumed that the magnetic properties of a superconductor could be deduced from its infinite conductivity, and it was only in 1933 that Meissner and Ochsenfeld first investigated the magnetic properties experimentally. They found that in fact some of the magnetic properties were entirely different from those to be expected on the basis of perfect conductivity. Their discovery stimulated a great deal of experimental and theoretical research, and though at first the situation was somewhat confused by various features of a secondary nature, such as complications due to impurities and shape, a clear and simple picture emerged after a few years, once the various effects had been sorted out. In this chapter, after contrasting the magnetic properties to be expected of a perfect conductor (§2.2) with the actual properties of superconductors (§2.3), we shall discuss the complication of specimen shape (§2.5) and introduce the idea of the 'intermediate state', but discussion of the nature of this state is deferred to Chapter IV. The behaviour of a ring (§ 2.6) and of other specimen shapes (§ 2.7) is next dealt with, followed by an account of the magnetic properties of superconducting alloys (§2.8). In §2.9 various problems of distribution of currents in superconductors (in the absence of applied magnetic fields) are discussed, and after a brief discussion of some other magnetic effects (§2.10), the chapter concludes with an account of experimental methods.

As has been mentioned in §1.4, the magnetic properties are seriously modified when the specimen size becomes comparable to the penetration depth ($\sim 10^{-5}$ cm.); in Chapter IV we shall see that in dealing with ellipsoidal specimens, size effects of a different character, associated with the structure of the intermediate state, occur even for much larger specimens. In this chapter we shall assume that we are always concerned with macroscopic specimens in the sense that the size is large enough for both kinds of size effect to be unimportant; in practice any size greater than a few milli-metres satisfies this condition. Except in § 2.8, which deals specific-ally with alloys, we shall also assume that the specimen is 'ideal' in the sense of having a perfectly sharp transition. Some remarks on the practical realization of this assumption will be found in the section on experimental methods (§ 2.11).

2.2. Magnetic properties of a perfect conductor*

The assumption of perfect conductivity is equivalent to assuming $E = 0$, and so we have at once from Faraday's law of induction that in the metal

$$\frac{1}{c}\frac{d\mathbf{B}}{dt} = -\operatorname{curl}\mathbf{E} = 0, \qquad (2.1)$$

and integrating, we find, if we assume unit permeability,

$$\mathbf{B} = \mathbf{H} = \mathbf{H}_0, \qquad (2.2)$$

where $\mathbf{H}_0$ is the field which was in the specimen when it lost its resistance. Thus as long as the specimen has perfect conductivity, the field distribution inside it cannot be changed by any external changes and can be regarded as 'frozen-in'. The physical meaning of this result is simply that any change of external magnetic field induces currents on the *surface* of the metal, and the magnetic field of these currents inside the metal just compensates the change of external field, thus keeping the field inside the metal constant. Since there is no resistance, such surface currents cannot die away, and so the field inside the metal remains constant with time.

We may notice that since

$$\mathbf{J} = \frac{c}{4\pi}\operatorname{curl}\mathbf{H},$$

any distribution of currents $\mathbf{J}_0$ in the body of the specimen at the moment it lost its resistance is likewise frozen-in, so that

$$\mathbf{J} = \mathbf{J}_0. \qquad (2.3)$$

If, for instance, a current is fed in from outside the metal, any *changes* in the current, after the metal has lost its resistance, flow as surface currents, leaving the body distribution unchanged.

* In this and later sections, vector symbols are used only where it is desirable to emphasize the vector nature of the quantity concerned.

The strength of the surface current **g** is given by the discontinuity of the tangential component of **H** in crossing the surface of the metal, i.e.

$$\mathbf{g}=\frac{c}{4\pi}[(\mathbf{H}_e-\mathbf{H}_0)\wedge\mathbf{n}],\qquad(2.4)$$

where **n** is the unit vector normal to the surface and $\mathbf{H}_e$ is the field just outside the surface. If we consider the case where no net current enters or leaves the specimen, the lines of **g** are closed and the magnetic moment of the specimen due to the surface currents

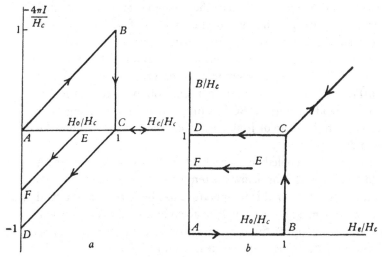

Fig. 5. Magnetic behaviour of perfectly conducting long cylinder.
(a) Magnetization curves. (b) B-H_e curves.

can be calculated by integrating the product of the current round each closed filament of the surface and the area of the filament, over all filaments. For the simple case of a long cylinder in a uniform field H_e parallel to its axis, this immediately gives for I, the magnetic moment per unit volume

$$I=-(H_e-H_0)/4\pi.\qquad(2.5)$$

Imagine now that we take this long cylinder through the sequences of operations indicated in fig. 5a or b. First let it lose its resistance in zero field, so that $H_0=0$, and then apply a field H_e. We shall then have

$$I=-H_e/4\pi,\quad B=0,\qquad(2.6)$$

and the specimen behaves as if it was diamagnetic with a susceptibility of $-1/4\pi$ per unit volume, or zero permeability. This continues until H_e reaches the critical value H_c, for restoration of resistance when the surface currents die out, and for higher value of H_e

$$I = 0, \quad B = H_e. \tag{2.7}$$

If now H_e is again reduced, the specimen loses its resistance in a field H_c, so for diminishing fields

$$I = -(H_e - H_c)/4\pi, \quad B = H_c. \tag{2.8}$$

These predicted I-H_e and B-H_e relations are shown graphically in figs. 5 a and b, and we see that there should be a very marked hysteresis effect, and that when the external field is zero, the cylinder should be left with a 'frozen-in' paramagnetic moment $H_c/4\pi$ corresponding to the 'frozen-in' flux, $B = H_c$. The physical meaning of this hysteresis is simply that for a reduction of the field the surface currents are induced in a sense opposite to that for an increase of field; the area $ABCD$ is proportional to the energy lost irreversibly in the form of Joule heat when the surface currents die away at B.

If the metal is cooled in some field H_0, no magnetic moment should be observed at or below the temperature at which it loses its resistance, but if, after this temperature has been passed, the field is reduced, the magnetization will be given by equation (2.5), which in this case again represents a paramagnetic moment (corresponding to the section EF in fig. 5 a). Correspondingly, B should have the value H_0 throughout and so not be affected by the loss of resistance.

2.3. Actual magnetic properties of a superconductor

Up to 1933 these predictions about the magnetic behaviour of a perfect conductor were assumed to apply to a superconductor, and this was regarded as so self-evident as not to require an experimental test. Thus in the early literature on superconductivity frequent references to the supposed 'frozen-in' moments can be found, although their existence had never been shown experimentally. In the course of some experiments on the magnetic field distribution round superconductors, Meissner and Ochsenfeld (1933) first showed that some of the predictions for a perfect conductor were quite wrong for an actual superconductor. They found, in fact, that for a pure superconductor, the field distribution always corresponded to zero field inside the superconductor, or in other words that inside a superconductor we always have

$$\mathbf{B} = 0, \qquad (2.9)$$

instead of $\mathbf{B} = \mathbf{H}_0$, independently of the initial conditions (i.e. of the field in which the metal became superconducting).

Later experiments showed that this result was quite general, and not connected in any way with the particular experimental arrangement used by Meissner and Ochsenfeld; let us see now what this result means in terms of our simple example of a long cylinder.

If the cylinder was cooled below the transition temperature in zero field, the magnetization will of course vary with a subsequent increase of field just as for a perfect conductor, i.e. $I = -H_e/4\pi$, and at the critical field the magnetization will as before disappear (this is simply because here $H_0 = 0$, and therefore $B = 0$ is equivalent to the result $B = H_0$ for a perfect conductor). It is only on again reducing the field that the properties of the actual superconductor begin to differ from the predictions for a perfect conductor; thus according to Meissner and Ochsenfeld's discovery no field can ever exist in a superconductor, and so when the field is reduced below H_c all the lines of force in the cylinder are suddenly pushed out, and the magnetization is again given by $I = -H_e/4\pi$, without any hysteresis. Similarly, if the metal is cooled in field H_0, as soon as the temperature reaches the critical value for this field, the lines of force are suddenly pushed out, and the cylinder acquires a magnetization $-H_0/4\pi$. In other words, the magnetization curve is always given by

$$I = -H_e/4\pi \qquad (2.10)$$

independently of the initial conditions, as shown in fig. 6a, and whatever the field in the body may have been just before it became superconducting, it is always zero in the superconducting state. Correspondingly, the B-H_e curve for the superconducting cylinder is as shown in fig. 6b, with B always zero in the superconducting state and $B = H_e$ in the normal state.

The absence of any magnetic flux in a superconductor, independently of the initial conditions, or the Meissner effect as it is called, is clearly an additional fundamental property of superconductors, since it cannot be deduced from the perfect conductivity. We shall see later (§6.2.2) that it can to some extent be regarded as the more fundamental property, and that both properties can be deduced from a single more general principle.

It should be noticed that if there is to be no magnetic field inside a superconductor, all currents must be superficial, and that if there was a body current originally flowing in the metal, this current becomes superficial immediately the metal becomes superconducting. Thus equations (2.3) and (2.4), which were predicted for a perfect conductor, must be replaced by

$$\left.\begin{array}{l} \mathbf{J} = \mathrm{o}, \\[1em] \mathbf{g} = \dfrac{c}{4\pi}\,[\mathbf{H}_e \wedge \mathbf{n}]. \end{array}\right\} \qquad (2.11)$$

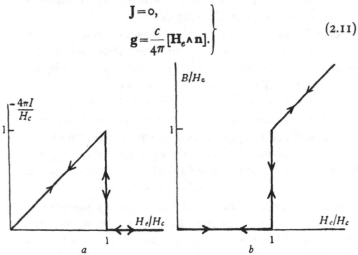

Fig. 6. Magnetic behaviour of superconducting long cylinder.
(a) Magnetization curve. (b) B-H_e curve.

In view of the theoretical importance of the Meissner effect, we shall anticipate slightly the fuller account of experimental methods given in §2.11, and describe briefly some simple experimental demonstrations of the results we have outlined so far. The simplest demonstration of the expulsion of all flux from a superconductor makes use of Faraday's law of induction. A coil is wound on a long cylinder of the metal which is to be made superconducting, and a magnetic field is switched on parallel to the axis of the cylinder, while the metal is still normal. A galvanometer connected in series with the coil then shows a ballistic deflexion proportional to the flux of the magnetic field in the metal. If now the specimen is slowly cooled, a ballistic deflexion opposite in sense, but equal in magnitude to the original one, is produced at the moment the transition temperature appropriate to the magnetic field is passed. This result and the fact that no further deflexion is obtained when the field is switched off show clearly that all the flux which entered into the metal in the normal state, is expelled when the metal becomes superconducting, and so provide the most striking demonstration of the distinction between the actual behaviour of a superconductor and the predicted behaviour of a metal characterized merely by perfect conductivity.

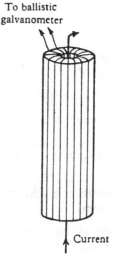

To ballistic galvanometer

Current

Fig. 7.

A somewhat similar experiment was carried out by Stark and Steiner (1937) to demonstrate the superficiality of current in the superconducting state. A coil connected to a galvanometer is wound round a hollow cylinder in the manner indicated in fig. 7, so that the lines of force of a current flowing along the *substance* of the cylinder are embraced by the coil. Just as in the previous experiment, equal and opposite ballistic deflexions are observed when a current is first switched on through the cylinder in the normal state, and when the cylinder is cooled through its transition temperature. This proves that all the lines of force of the current inside the metal are ejected when it becomes superconducting, and therefore that no current remains in the body of the metal. Other experimental proofs of the superficiality of current are mentioned on pp. 30 and 49.

Finally, we may mention one other demonstration of the zero permeability, which is remarkable mainly for its aesthetic appeal. This is the floating-magnet experiment first carried out by Arkadiev (1945, 1947). If a small permanent magnet is lowered over a superconducting surface, the lines of force of the magnet cannot enter the superconductor, and this produces a repulsive force large enough to overcome the weight of the magnet. An actual photograph of a magnet floating in this way is shown in Plate I. The simplest way of understanding the repulsion is to think of image poles under the superconductor (see fig. 8) which produce in the space above the superconductor just the correct field distribution appropriate to the zero permeability of the metal. Alternatively, we may think of the repulsion as just the mutual repulsion between a magnet and a diamagnetic body.

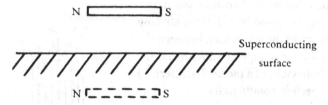

Fig. 8. Illustrating the principle of the floating magnet experiment. The real magnet and its image would, in the absence of the superconductor, produce above the surface just the same field distribution as produced by the real magnet alone and the superconductor.

2.4. Some remarks on the formal description of magnetic properties

Before going any further, some discussion is necessary of the precise meaning of our magnetic vectors. There has, in fact, been some inconsistency in speaking of a superconductor as having zero permeability when we have earlier supposed that $B = H$. The confusion arises because there are really two different ways of describing the magnetic properties.

In the first way (which has been used implicitly up to now) $B = H$ and the currents appear explicitly as g or J, just as would the eddy currents in an alternating field problem, though with the difference that here the frequency is zero and the currents are time invariant. The magnetic moment for unit volume, I, is merely a quantity derived from the current distribution, and has not the meaning of $(B - H)/4\pi$. In this type of description it will be noticed that there is no distinction between B and H, which both vanish inside a superconductor, so that either may be used alone to describe the 'field', and there is no distinction between currents fed in from outside and currents induced by the applied magnetic field. It is this type of description which is used in the Londons' theory (§ 6.2). An alternative description, which is very convenient for many purposes, and leads always to identical results if properly interpreted, regards the superconductor as really having zero

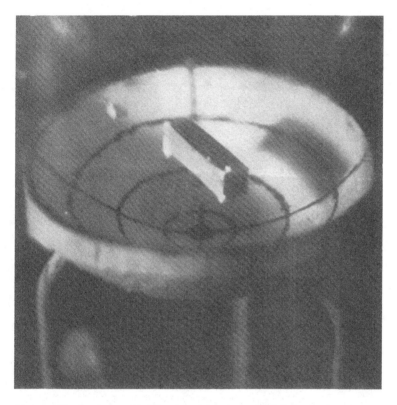

Plate 1. *The floating magnet.* The short bar magnet (slightly larger than life size) is floating in helium gas nearly half an inch above the bottom of the superconducting lead bowl; the shadow of the magnet is visible on the right-hand side of the bowl. The bowl is painted white with black lines on it to bring out the perspective and it is standing on three copper legs which dip into liquid helium to keep it superconducting (the liquid helium is not visible in the picture). The white specks on the magnet and in the bowl are small pieces of solid air (the air is a slight impurity in the helium).

permeability. In this description $\mathbf{H}$ is defined differently, so that

$$\mathbf{B} = \mathbf{H} + 4\pi\mathbf{I}, \qquad (2.12)$$

and $\mathbf{I}$ is regarded as a genuine magnetization of the body. The currents associated with an applied field now no longer appear explicitly, and it is only currents fed in from outside (or circulating currents in a ring, see § 2.6) which are taken into account explicitly in the equations

$$\mathbf{J} = \frac{c}{4\pi}\operatorname{curl}\mathbf{H} \quad \text{and} \quad \mathbf{g} = \frac{c}{4\pi}[(\mathbf{H}_1 - \mathbf{H}_2)\wedge\mathbf{n}], \qquad (2.13)$$

where now $\mathbf{H}_1$ and $\mathbf{H}_2$ are the values of $\mathbf{H}$ immediately outside and inside the surface. Thus when there are no such 'external' currents, the tangential component of $\mathbf{H}$ is continuous across the surface and, outside the superconductor, $\mathbf{H}$ is, as before, just the ordinary magnetic field satisfying the boundary conditions of the problem.

In order to make clear the difference between the two approaches, we shall write down the two descriptions for the long superconducting cylinder (radius a), in a uniform field $\mathbf{H}_e$ parallel to its axis (z direction), both with and without a current i fed into the cylinder. In the first description we have ($\mathbf{n}$ and $\mathbf{z}$ are unit vectors)

	Without current	With current i
Outside	$\mathbf{B}=\mathbf{H}=\mathbf{H}_e$	$\mathbf{B}=\mathbf{H}=\mathbf{H}_e+\dfrac{2i}{cr}[\mathbf{n}\wedge\mathbf{z}]$
Inside	$\mathbf{B}=\mathbf{H}=0$	$\mathbf{B}=\mathbf{H}=0$
Surface	$\mathbf{g}=\dfrac{c}{4\pi}[\mathbf{H}_e\wedge\mathbf{n}]$	$\mathbf{g}=\dfrac{c}{4\pi}[\mathbf{H}\wedge\mathbf{n}]$

In the second description we have

	Without current	With current i
Outside	$\mathbf{B}=\mathbf{H}=\mathbf{H}_e$	$\mathbf{B}=\mathbf{H}=\mathbf{H}_e+\dfrac{2i}{cr}[\mathbf{n}\wedge\mathbf{z}]$
Inside	$\mathbf{B}=0,\ \mathbf{H}=\mathbf{H}_e,\ \mathbf{I}=-\mathbf{H}/4\pi$	$\mathbf{B}=0,\ \mathbf{H}=\mathbf{H}_e,\ \mathbf{I}=-\mathbf{H}/4\pi$
Surface	$\mathbf{g}=0$	$\mathbf{g}=\dfrac{i\mathbf{z}}{2\pi r}$

For shapes other than the long rod, the solution is slightly more complicated, and it is here that the second description is particularly advantageous, since if the superconductor is treated as a body of zero permeability it is only necessary to apply the standard solution of the problem of a body of given permeability in an applied magnetic field. Before discussing this in the next section, we must say a word about the meaning of 'applied' field, which we shall always denote by H_e. We shall always suppose that our magnetic fields are produced by iron-free coils, and H_e then simply means the field at any point before the specimen was first there (if iron is used in the production of the field the situation is complicated by inter-actions between the specimen and the iron, so that H_e cannot be defined in such a simple way). For the long cylinder in a uniform field H and H_e are identical outside the cylinder, but this is no longer true for any other shape, when the field H outside is made up of the original field H_e together with the field produced by the currents in the specimen (or in terms of the second description, by the diamagnetism of the specimen).

2.5. Magnetic properties of a superconducting ellipsoid in a uniform magnetic field

Following the approach indicated at the end of § 2.4, we shall now consider the magnetic properties of a superconducting ellipsoid in a uniform field H_e, treating the superconductor as a body of zero permeability. We shall suppose that H_e is parallel to a principal axis of the ellipsoid, but the more general case is easily deduced by resolving H_e along each axis in turn and using the superposition principle. If the demagnetizing coefficient is $4\pi n$ and the perme-ability is μ, the general solution* for the field distribution is given by the equations

Inside:
$$\left.\begin{aligned} B &= \mu H, \\ H &= H_e - n(B-H), \\ I &= (B-H)/4\pi; \end{aligned}\right\} \tag{2.14}$$

Outside:
$$B = H = H_e + \operatorname{grad} \phi, \tag{2.15}$$

where ϕ is the solution of $\nabla^2 \phi = 0$ appropriate to a uniform distribu-tion of magnetization I over the ellipsoid. Fortunately, we shall not usually need more than a limited amount of information about the field outside, so that the detailed form of ϕ need not be given.†

---

* See for instance, J. C. Maxwell, 1892, Vol. II, p. 66; for values of n, see Stoner, 1945.

† For a sphere of radius a, ϕ has the form

$$\phi = \frac{\mu-1}{\mu+2}\frac{a^3}{r^3}(H_e . r).$$

For the special case of $\mu = 0$, the solution inside the specimen reduces to

$$\left.\begin{array}{l} B = 0, \\ H = H_e/(1-n), \\ I = -H_e/4\pi(1-n). \end{array}\right\} \qquad (2.16)$$

Thus the slope of the magnetization curve of a superconducting ellipsoid is the steeper, the greater its demagnetizing coefficient, i.e. the flatter it is transversely to the field direction. For a long cylinder in the field direction, $n = 0$, and (2.16) reproduces the result already obtained in §2.3 (2.10). Other special cases of practical importance are a cylinder with its axis perpendicular to the field, when $n = \frac{1}{2}$, so that $I = -H_e/2\pi$; a sphere, for which $n = \frac{1}{3}$ and $I = -3H_e/8\pi$, and a flat disk, with its plane normal to the field, for which $n = 1$, and I increases infinitely rapidly (in this case, however, equation (2.16) is valid over only an infinitely small range of fields, since, as we shall see directly, the field at the edge of the disk immediately exceeds H_c). We may point out that the result (2.16) could equally well have been obtained from the description of infinite conductivity if the ellipsoid became superconducting in zero field, for as we saw in §2.2, this case is equivalent to $B = 0$; the infinite conductivity description would make a different prediction, however, if the specimen had become superconducting in a non-zero field, while the Meissner effect shows that (2.16) must be valid independently of the initial conditions.

As regards the field outside the specimen, all we shall need are the following features: (a) **H** is tangential at the outside surface (since its normal component equals the normal component of **B** and must therefore vanish), (b) **H** falls off away from the surface, (c) the tangential component of **H** has to be continuous across the boundary, so if θ is the angle between the normal and **H**$_e$ then

$$H = H_e \sin \theta/(1-n)$$

at any point just outside the surface. Clearly the largest value of H is in the 'equatorial plane' where the surface is parallel to **H**$_e$, and this maximum value is given by

$$H_{max} = H_e/(1-n).$$

A sketch of the lines of force of an originally uniform magnetic field applied transversely to a long circular cylinder is shown in fig. 9.

The most important consequence of these features is that the field just outside the surface will reach the critical value H_c for destruction of superconductivity sooner at some places than at others. Indeed, at the 'equator' of the ellipsoid the field is $1/(1-n)$ times H_e, and so we should expect the field there to exceed H_c, and destruction of superconductivity to begin, as soon as H_e exceeds $(1-n)H_c$.

Fig. 9. Lines of force round a superconducting cylinder in a transverse field. The lines have the equations $(r^2-a^2)\sin\theta=cr$ in polar co-ordinates, where a is the radius of the cylinder and c has the values 0, $\pm\frac{1}{2}a$, $\pm a$, $\pm\frac{3}{2}a$, $\pm 2a$, $\pm\frac{5}{2}a$ for the various lines shown (von Laue, 1949).

As H_e is increased beyond this value, it might at first sight seem that the destruction of superconductivity would take place by a progressively larger region round the equator reverting to the normal state, with some boundary surface separating it from an inner still superconducting region, until when $H_e=H_c$, the inner region was completely 'eaten away', and the whole specimen was in the normal state; this transition mechanism, however, leads immediately to difficulties. Suppose, for in

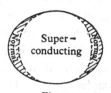

Superconducting

Fig. 10.

stance, that at some stage of the transition there is a boundary surface, as shown in fig. 10, between an inner superconducting region and an outer normal one. Evidently this boundary will be determined by the condition that the field is just equal to H_c on it (or is less than H_c where the boundary coincides with the surface of the specimen); the field, however, outside such a convex* superconducting region will be less than at its surface (just as for an ellipsoid, as already pointed out), and consequently in the region between the boundary surface and the surface of the ellipsoid the field is less than H_c, and there is no reason why this

* It can easily be seen that some part of the boundary will have to be convex.

part should be in the normal state. It is, in fact, impossible to find a simple boundary surface which satisfies simultaneously the conditions that the field should have the value H_c over it, and that the field should be greater than H_c in the region between it and the surface of the ellipsoid. This means that when $H_e > (1 - n) H_c$, the ellipsoid must split up into some arrangement of superconducting and normal regions, more complicated than that suggested in fig. 10, and we shall see in Chapter IV that the division is on a fine scale.

Peierls (1936) and F. London (1936) have shown, however, that the macroscopic magnetic properties of the ellipsoid, when it is split up in this way, i.e. when

$$H_e > (1 - n) H_c,$$

can be obtained without any detailed knowledge of the structure of this fine division, by assuming that on a macroscopic scale the specimen is in a uniform 'intermediate state' (the uniformity is justified by experiment), with B intermediate between 0 and H_c, and $H = H_c$ (it will be remembered that this is the value of H just before commencement, and just after termination of the intermediate state).

If we assume then, that for $H_c > H_e > (1 - n) H_c$, the specimen is in a macroscopically uniform intermediate state with $H = H_c$, we find at once from (2.14) and (2.15)

$$\left. \begin{array}{l} B = H_c - (H_c - H_e)/n \\ I = (B - H)/4\pi = (H_e - H_c)/4\pi n. \end{array} \right\} \quad (2.17)$$

and

These equations are evidently reasonable, since they give correct results at both ends of the intermediate state range. Thus for $H_e = (1 - n) H_c$, $B = 0$ and $I = -H_c/4\pi$ in agreement with (2.16), while for $H_e = H_c$, $B = H_c$ and $I = 0$ corresponding to full destruction of superconductivity. It should be noted that the area under the I-H_e curve is always $H_c^2/8\pi$, whatever the value of n. We shall see in § 3.1 that this is a thermodynamic requirement, and the fact that it comes out of the present treatment indicates that the treatment is consistent with thermodynamics.

The full B-H_e, H-H_e and I-H_e curves for a sphere $(n = \frac{1}{3})$ covering the superconducting, intermediate and normal state ranges are illustrated in fig. 11. In fact, curves very much like these were first found experimentally, and the theory just outlined was developed to explain them. Mendelssohn and Babbitt (1935) first studied the field at the equator of a tin sphere by the bismuth wire technique (see §2.11), and a more detailed survey by the same method was later made by de Haas and Guinau (1936). The field at the equator should of course be just H, while the field at a pole should be B, and apart from slight irreversibilities the curves of fig. 11 a were indeed found. Measurements in a thin canal parallel to the field direction

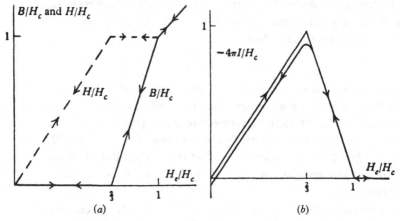

Fig. 11. Magnetic behaviour of a superconducting sphere. (a) B-H_e and H-H_e curves (theoretical, but in good agreement with de Haas and Guinau (1936) for tin). (b) I-H_e curve for lead (Shoenberg, 1936).

showed directly that H was indeed H_c in the intermediate state, but in the superconducting state the field stayed at the value it had when superconductivity was fully established, i.e. on increasing H_e from zero the field in the canal was zero and on reducing from above H_c, the field stayed at H_c. Indeed, in the superconducting state H is really little more than a mathematical fiction, and does not represent any real field in the metal. If a canal is bored right through the superconductor, the specimen is no longer simply connected and fields are frozen-in as in a ring (see §2.6), so that the theoretical H is not measured. Even if it were possible to construct an internal hole with no open ends, no field would be found in it for increasing H_c, because the uniform distribution of H characteristic of a perfect ellipsoid would always be upset by the presence of the hole in such a way as to reduce H to zero at the boundary of the hole; for reducing H_e from above H_c frozen-in field effects would occur, as is usual in specimens of non-ellipsoidal shape (see §2.7). Measurements of I for a sphere were first made by Shoenberg (1935, 1936), and again good agreement with theory was obtained. The slight hysteresis indicated in fig. 11 b was probably due to slight impurities in the lead sphere.

2.6. The superconducting ring

A multiply-connected body, such as a ring, differs essentially from a singly-connected one in that it is possible to induce a 'total' current round it, in the sense that closed circuits can be drawn enclosing parts of the body (but nowhere passing through it) which enclose a net flow of current. If the body concerned is not super-conducting, any such total current is always associated with an electric field, so that Joule heat is produced, and the current cannot maintain itself; in a superconductor, however, the current is not associated with any electric field, so that it can flow indefinitely.

To illustrate the peculiar magnetic properties caused by the possibility of inducing such 'persistent' total currents, we shall consider a case which can be treated mathematically, that of a circular ring* of mean radius b, made of a wire of circular cross-section and radius a, small compared with b. If the self-inductance of the ring is L, then any change of a magnetic field H_e normal to the plane of the ring will cause a change of the total current i round the ring, given by
$$L\,di/dt - \pi b^2\,dH_e/dt = 0,$$

which expresses the fact that there is no electromotive force in the superconducting ring. Integrating this, we find

$$Li = \pi b^2(H_e - H_0), \qquad (2.18)$$

where H_0 is the value the field had when there was no current in the ring. If there were no current flowing, the ring would have just the magnetic moment of a long wire in a transverse field (to a first approximation the fact that the wire is bent into a circle is irrelevant), i.e. $-\pi ba^2 H_e$ would be the moment in field H_e. The total current i,

---

* The results which do not involve the circular symmetry of the ring apply to any type of ring; other results are, however, only qualitatively applicable to non-circular rings.

however, gives the ring an additional magnetic moment $-\pi b^2 i$, which is of the order of magnitude $(b/a)^2$ times larger. This can be seen by substituting the value of L from (2.20) below, but it is qualitatively evident, since the moment due to the total current is of the order of magnitude of the moment of a sphere of radius b, while the moment of the ring when there is no current is proportional only to the volume of the superconducting material in the ring. This division of the magnetic moment into two parts is somewhat artificial (see § 2.4), but it is a convenient way of describing the real current distribution, which consists of a surface current unevenly spread over the cross-section of the wire. The current i represents the mean current flowing in one direction, while the currents responsible for the magnetic moment when $i = 0$ are those which remain when i has been subtracted; roughly speaking these are equal currents in opposite directions, flowing along the inside and outside surfaces of the ring. It is because these equal and opposite currents flow so close to each other that their magnetic moment is so small relative to that of the net current i in one direction.

We see at once from (2.18) that the magnetic properties of a ring differ essentially from those of a singly-connected body, since the value of the magnetic moment depends entirely on the initial conditions. If, for instance, the ring was cooled below its transition temperature in zero field, $H_0 = 0$, and the current is given by

$$Li = \pi b^2 H_e, \tag{2.19}$$

while if it was cooled in field H_0, the current is given by (2.18), and in particular when the field is reduced to zero after the cooling, a current $-\pi b^2 H_0/L$, corresponding to a large 'paramagnetic' moment, is left flowing in the ring. This lack of uniqueness of the magnetic moment is very similar to that which we deduced for a metal of infinite conductivity, and arises from the fact that the current round the ring is induced in such a way as to keep the flux through the hollow of the ring constant. The difference between a singly-connected superconductor and a superconducting ring is that in the former the permeability is zero all over any cross-section, and whatever flux passed through it originally is always completely pushed out when it becomes superconducting. In the latter, however, the permeability of most of the cross-section (i.e. the hollow) remains unity, so that a field H_0 can remain in the hollow of the ring even when the ring has become superconducting.

The constancy of the current in a superconducting ring offers the most accurate confirmation of the effectively zero resistance of the metal in the superconducting state. If in fact there was any electromotive force associated with the current, energy would be lost in the form of Joule heat, and the current would die away in accordance with the relation $L\,di/dt + Ri = 0,$

R being the resistance of the ring (i.e. the ratio of the e.m.f. to the current); we should then have

$$i = i_0\,e^{-Rt/L}.$$

The time of decay of the current, $\tau = L/R$, for a metal at low temperature but not superconducting is, for any practical dimensions of the ring, fairly short, usually not more than a few seconds; for a superconductor, however, even with a ring of very thin wire or a thin film coated on a non-superconducting wire (to increase the value of the hypothetical R), no decay of the current could be observed over several hours, and in this way, from the limit of change of current which the experimental method used could detect, an upper limit to any possible value of R could be deduced (see p. 6).

Returning to the discussion of the magnetic properties of a superconducting ring, we shall now describe the experimental results. Shoenberg (1936) investigated the magnetic properties by measuring the total magnetic moment of the ring with the force method (this method has the disadvantage that only rather small rings can be used, and consequently it is very difficult to make the ring sufficiently uniform in cross-section for an exact comparison with the theory), while Smith and Wilhelm (1936) and Schubnikow and Chotkewitsch (1936) measured the field at various places round the ring, and thus deduced the magnetic properties. In fig. 12 we show how the magnetic moment of a ring which was cooled in zero magnetic field varies with subsequent increases and decreases of a field normal to the plane of the ring (actually this diagram is a slightly idealized version of the experimental results, which in some details are confused by other factors which are of no interest here). Initially the magnetic moment increases with the

field just as explained above; when the moment of the ring, in the
absence of a total current, has been subtracted, the slope of the line
OA agrees very well with (2.19), and it is interesting to mention that
this agreement confirms the fact that the current flows only over the

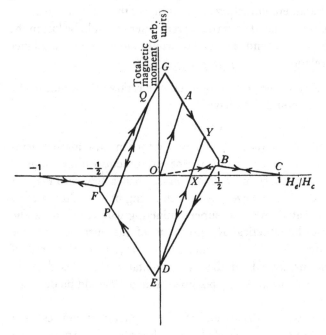

Fig. 12. Magnetization curve of a superconducting ring (schematic).

surface of the ring. Thus the experiments show that (2.19) is verified
quantitatively only if for L we put

$$L = 4\pi b(\ln 8b/a - 2), \qquad (2.20)$$

the value it assumes for an entirely superficial current, and which
differs by about 10 %, for the actual rings used, from the value

$$L = 4\pi b(\ln 8b/a - \tfrac{7}{4})$$

of the self-inductance for a current flowing through the whole cross-
section of the ring.*

* The difference between these two values of L was also shown by a.c.
measurements at 50 cyc./sec. (Shoenberg, 1937c), but from these it cannot of
course be deduced that the current would be superficial for zero frequency—
the superficiality of an a.c. current merely points to a very complete skin effect.

As soon as the point A is reached, (2.19) ceases to be valid, and the current begins to fall off linearly with further increase of field. The reason for this falling off is that the total field at some part of the surface of the ring (in this case the outer rim) reaches the value H_c when the external field exceeds a certain value which we shall calculate directly. As soon as this happens, resistance reappears in the ring and the current round the ring dies away to a value which will keep the maximum field at the surface just critical, so that the ring becomes superconducting once more. Thus in the section AB of the magnetization curve, the current in the ring decreases with increasing field in such a way as to prevent the field anywhere at the surface of the ring exceeding H_c; the ring is, however, completely superconducting throughout the section OAB, as is shown by the fact that a decrease of field from any point Y on AB causes a change of current in agreement with (2.18) with the superficial value of L, i.e. along a line parallel to OA. An a.c. measurement of the inductance (Shoenberg, 1937c) also showed that the ring was completely superconducting in the region AB as well as OA.*

It is easily seen that the field at the surface of the ring has a maximum value at the *outer* rim of the ring in the section AB of the magnetization curve, for there, the field H_I of the total current, and the field H_E, which would be there in the absence of a total current, are in the same direction. If we neglect correction terms of order a/b,† we have $H_I = 2i/a$ and $H_E = 2H_e$, so that the field corresponding to the point A is given by

$$2H_e + 2i/a = H_c, \qquad (2.21)$$

or substituting from (2.19) for i, we have

$$H_e = \tfrac{1}{2}H_c/(1 + \pi b^2/La), \qquad (2.22)$$

for the field at which the current begins to fall off.

Beyond A, as we have already pointed out, the current can no longer assume its 'full' value $\pi b^2 H_e/L$, but adjusts itself to satisfy

---

* The interpretation of the region AB of the magnetization curve originally given by Shoenberg (1936) is definitely wrong.

† A better approximation has been given by Schubnikow and Chotkewitsch (1936), but we do not require this for our qualitative considerations. We shall not, however, neglect terms such as $La/\pi b^2$ in comparison with unity, since although roughly proportional to a/b, the proportionality factor is large compared with unity and increases with decrease of a.

(2.21), or, in other words, the current along the section BC of the magnetization curve is given approximately by

$$i = a(\tfrac{1}{2}H_c - H_e).$$

We see, then, that as soon as the external field H_e reaches the value $\tfrac{1}{2}H_c$, there will no longer be any current flowing round the ring (the better approximation shows that at $\tfrac{1}{2}H_c$ there is still a small current, which disappears discontinuously for further increase of field), and the ring at the same time enters the intermediate state. For further increase of field, the ring behaves almost exactly in the same way as a long wire in a transverse field greater than $\tfrac{1}{2}H_c$ (see §2.5), since there can no longer be any total current round the ring (any total current, however small, would make the field at the surface of the ring exceed H_c, and, moreover, owing to the resistance of the metal in the intermediate state, such a current would in any case die away). In fact, in this region the multiple connexion of the ring is no longer of any importance. This was confirmed experimentally by repeating the measurements after the ring had been cut; the experiment showed that the same magnetization was obtained over the range BC whether the ring was cut or not, although of course the magnetization in the range OAB was greatly reduced by cutting the ring (OB refers to the cut ring).

As soon as the external field reaches H_c, superconductivity is completely destroyed and no magnetic moment is left; if now the field is again reduced, the portion BC in fig. 12 is retraced, no total current being induced in the ring. At $H_e = \tfrac{1}{2}H_c$, however, the ring becomes completely superconducting again, and further decrease of field induces a current in the ring in a sense *opposite* to that of the current originally induced by an increase of field. Actually the better approximation shows that this happens at a field slightly less than $\tfrac{1}{2}H_c$, in fact at

$$H_e = \tfrac{1}{2}H_c/(1 + a/4b).$$

We might thus expect that below $\tfrac{1}{2}H_c$ the current would be given by (2.18) with $H_0 = \tfrac{1}{2}H_c$, since the ring had zero current for this field. This current, induced on reducing the field, can, however, never attain its full value as given by (2.18), for if it were to do so, the field would exceed H_c at the surface of the ring. The field at the surface of the ring is now greatest at the *inside* rim of the ring (since the current has now the opposite sense), and the total field there is of magnitude $2H_e - 2i/a$. Thus if i were to attain its full value

$$(\pi b^2/L)(H_e - \tfrac{1}{2}H_c),$$

this total field would be $(\pi b^2/La)H_c - 2H_e(\pi b^2/La - 1)$, which is, in fact, always greater than H_c if, as here, $H_e < \tfrac{1}{2}H_c$ (it is easy to show that the dimensionless ratio $\pi b^2/La$ is always greater than unity). Since the current cannot attain its full value, it will, as along AB, be given by the condition that the total field at the surface of the ring (this time at the inner rim) is just H_c; in other words, the current along the portion BD of fig. 12 will be given by

$$i = -a(\tfrac{1}{2}H_c - H_e).$$

It should be noticed that the slopes of AB and BD are just equal but of opposite sign (this again is true only to our approximation, and if the magnetization not due to the total current is neglected). Exactly similar considerations show that the portion EF is determined by the condition that the field should be just critical on the outer rim; FG, by the condition that it should be critical on the inner rim; and for GAB again on the outer rim. We see that the greatest 'persistent' current that can be left in the ring in zero field is given by $2i/a = H_c$, and can be made to flow in either sense round the ring. It may at first sight seem puzzling why the linear portion BD continues to E, for it would seem that as soon as the field was reversed, the greatest magnitude should occur at the outer rim again, where H_I and H_E have the same direction; actually, however, the better approximation shows that the magnitudes of H_I at the inner and outer rims differ by a small quantity proportional to a/b (and similarly for H_E), so that the field continues to be maximum at the inner rim for a little way after the field has been reversed (in fact until E). Similar considerations explain the position of the turning-point G.

The line $ABDEFG$ can be considered as a kind of boundary curve limiting the possible values that the total current in the ring can assume—anywhere inside this boundary the current varies with a change of field according to (2.18), e.g. along lines such as DX or PQ, parallel to OA, but as soon as the current reaches a point on the boundary curve, any further change of field in the same direction alters the current along the boundary curve itself. This consideration shows that to produce the maximum 'persistent' current in a ring in zero field, it is necessary to cool the ring only in the field corresponding to the point X in fig. 12 and then to remove this field; if the ring had been cooled in zero field, it would be necessary to put on and remove a field at least as large as that corresponding to Y.

For completeness it is useful to record here the equations for the boundary curve obtained with the better approximation of Schubnikow and Chotkewitsch (1936). These are

$$i = \mp a\{\tfrac{1}{2}H_c - H_c(1 \mp \alpha)\}/(1 \pm \beta),$$

where $\alpha = a/4b$ and $\beta = 2\alpha(1 + \ln 2/\alpha)$; the lower signs are to be taken when the field is critical at the outer rim, and the upper signs when the field is critical at the inner rim (H_c must be taken negative for the sections in the left-hand part of the diagram). The observed magnetic moment will of course also contain a term due to the magnetic properties of the ring with no total current (p. 27). When α is no longer small compared with unity no useful explicit formulae can be given, but numerical computations for this general torus have been made by de Launay (1949) and verified experimentally by Dolecek and de Launay (1949).

2.7. Miscellaneous non-ellipsoidal shapes

In general, no mathematical treatment can be given of the magnetic behaviour of specimens whose shape is not at least approximately ellipsoidal, but certain features of the behaviour of such specimens are of practical interest, and we shall give a brief account of the various experimental investigations which have been made.

The behaviour of short cylinders of various length to diameter ratios has been studied by Shoenberg (1937a), and some typical magnetization curves for the field parallel and perpendicular to the cylinder axis are shown in fig. 13. For both settings, the curve no longer turns over quite so sharply as for an ellipsoid; this can be interpreted as due to an inhomogeneous distribution of H in the cylinder. Thus probably there is no longer a macroscopically uniform intermediate state during the destruction of superconductivity, but some more complicated mixture of superconducting and normal regions, which varies in composition through the specimens. For the parallel setting, when the field is reduced below H_c there is very marked hysteresis rather reminiscent of that found for a ring (fig. 12). The probable explanation of this hysteresis is that, owing to the inhomogeneous field distribution, a ring on the surface of the specimen is able to become superconducting earlier than the rest of the specimen as the field is being reduced. For further reduction of the external field a current is induced in this ring, thus keeping the flux constant in the interior regions and preventing them from becoming superconducting again.

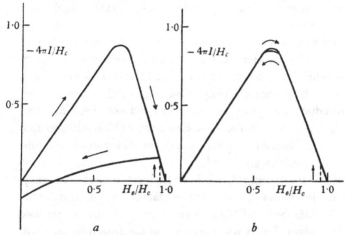

Fig. 13. Magnetization curves of a short tin cylinder (Shoenberg, 1937a); I is the *average* magnetization. (a) Field parallel to cylinder axis. (b) Field perpendicular to cylinder axis. The discontinuous reappearance of magnetization (indicated by the broken lines) is discussed on p. 121.

Though the formation of such rings has not yet been deduced from theoretical knowledge of the field distribution, but has merely been postulated *ad hoc* to explain the hysteresis, there is some other experimental evidence which supports this qualitative explanation. It was found that the hysteresis of fig. 13a could be considerably reduced by rounding off the rims of the cylinder, which probably made the field distribution more homogeneous, and in terms of our explanation thus delayed the formation of a superconducting ring. Moreover, there was practically no hysteresis at all if the field was perpendicular, instead of parallel, to the axis of the cylinder (fig. 13b); in this case there is no symmetry round the field direction, and so it is impossible for the specimen to become superconducting all round a section normal to the field simultaneously.

The curves of fig. 13, and in particular the form of the hysteresis, were found to be practically independent of temperature, i.e. the magnetization curves at different temperatures could be brought into coincidence by a mere change of scale. This temperature independence provides a practical means of distinguishing the hysteresis due to shape from that due to impurities (see § 2.8), which is usually temperature-dependent.

Somewhat similar hysteresis effects have been found in flat plates transverse to an applied field by Alekseyevsky (1941 a, 1946), who has studied in detail the topography of the frozen-in flux, and by Andrew and Lock (1950) (other aspects of whose experiments will be discussed in § 4.5). Here, presumably, the hysteresis was again due to the difficulty of suitably rounding off the plate edges, and it was found, indeed, that electrolytic etching which smoothed the edges did reduce the hysteresis. Another specimen shape that has been studied in some detail is the hollow sphere (Mendelssohn and Babbitt, 1935; Shalnikov, 1940b, 1942), for which again complicated hysteresis effects are found.

It is possible that a somewhat different hysteresis mechanism may occasionally be relevant. In some measurement on transverse cylinders by Désirant and Shoenberg (1948a), the hysteresis was usually very slight, but it was noticed that the frozen-in moment could be made much larger if the field was switched off suddenly from a high value instead of being reduced gradually (as was the usual procedure). This suggests that if the field is reduced too rapidly there is not sufficient time for the flux to get out of some parts of the specimen, and it is trapped in normal regions by closed superconducting circuits surrounding them. This interpretation is supported by the fact that time-lag effects have been observed in the intermediate state region.* Thus de Haas, Engelkes and Guinau (1937) found that after a sudden change of applied field, the field distribution in a sphere assumes its final form only gradually (the time lag was of order 15 min. at $\frac{2}{3}H_c$ and negligible near H_c). Mendelssohn and Pontius (1936a, b) have found that such

---

* The time effects occurring in the transition of a long cylinder to the super-conducting state after 'supercooling' are discussed in § 4.6.3.

effects are particularly marked in specimens of non-ellipsoidal shape.

We have up to now discussed only the magnetic properties of specimens which have become superconducting by cooling in zero external field. If the metal is cooled in a non-zero field, no flux is frozen-in if the specimen is a perfect ellipsoid (i.e. entirely free from blemishes), and so all the results of § 2.5 are unaffected. If, however, the specimen is non-ellipsoidal or contains mechanical flaws, super-conducting rings may be formed which still enclose normal regions, and the lines of force passing through these normal regions are then trapped, leaving eventually a frozen-in flux at the final temperature. Such freezing-in is sometimes important if it causes undesirable effects in delicate experiments (see, for instance, § 5.3.1), and special measures must then be taken to compensate the earth's magnetic field during cooling of the specimen.

2.8. Magnetic properties of alloys

Apart from the pure element superconductors whose properties we have been discussing up to now, many alloys are known to become superconducting at low temperatures (see Appendix I, Tables X and XI). Roughly speaking it is possible to divide these alloys into two classes:

(1) Alloys of the superconducting elements either with each other, or with elements which are not at present known to become superconducting and which can become superconducting in a large range of concentrations; in these the superconducting property cannot be said to be characteristic of any particular composition corresponding, for instance, to a chemical compound. The transition temperatures are of much the same order of magnitude as for simple superconductors, but the transition is spread over a relatively large temperature range (of order $1°$ K.). In the literature it is usual to give the transition temperature as that for which the resistance has half its full value, but it should be remembered that this temperature is in general sensitive to the strength of the measuring current. There is no universal rule as to how the transition temperature depends on the composition, but often (particularly for addition of bismuth) the transition temperature

of the pure superconducting element is increased by the addition of a non-superconducting component.

(2) Chemical compounds between certain elements (both super-conducting and non-superconducting); as examples of this type we may mention Au_2Bi (de Haas, van Aubel and Voogd, 1929 a, c),* CuS (Meissner, 1929 a)† and other bismuth compounds recently studied by Alekseyevsky (1945 c, 1948 b, 1949 b, 1950). The experi-mental evidence suggests that alloys of this type would, if pre-pared pure (i.e. with exactly the right composition throughout the specimen), behave like pure element superconductors. Thus the critical field values for Au_2Bi, BiNa, and CuS are of the same order of magnitude as for a pure metal rather than an alloy of class (1), and it has also been shown (Shoenberg, 1938) that if Au_2Bi is suitably prepared its magnetic properties do not differ appreciably from those of a pure element superconductor (i.e. complete Meissner effect). It seems in fact as if in these alloys the super-conductive property is characteristic of the particular arrangement of the atoms corresponding to the lattice of the chemical compound. It is interesting to notice that for all superconducting alloys where neither component is alone superconducting, at least one of the components lies immediately next to one of the two groups of superconducting elements in the periodic system (see Appendix, Tables IX and X). This rather confirms the suggestion (§ 1.1) that the grouping in the periodic system is not accidental.

In this section we shall be concerned mostly with the properties of the alloys of the first category, which differ in almost every essential respect from those of pure element superconductors.‡ The existence of the other class of alloys may prove to be of importance in the development of a theory of superconductivity, but since it is probable that they do not differ from the simple superconductors in their general properties, we shall not need to

---

* See also de Haas and Jurriaanse (1932), Jurriaanse (1935).

† See also Buckel and Hilsch (1950).

‡ Recent work has shown that this is not always so. Lasarew and Nakhutin (1942) found that alloys of tin and zinc (which are mutually insoluble, i.e. mixtures rather than solid solutions) have critical fields very similar to those of pure tin, while Stout and Guttman (1950) found that alloys of indium and thallium, if prepared as single crystals, behave very much like pure elements (i.e. fairly complete Meissner effect and normal critical fields) in spite of being disordered solid solutions.

discuss them specially. It should therefore be remembered that when in future we use the word alloy, we shall always have in mind alloys of the first kind. It may be mentioned here that Lasarew and Galkin (1944) have shown that strong inhomogeneous mechanical deformation of a pure metal also causes most of the effects that are to be described below as characteristic of alloys.

The first characteristic difference found between an alloy and a pure superconductor was in the restoration of the resistance by a magnetic field (see, for instance, de Haas and Voogd, 1930, 1931 b). The resistance field curve for an alloy is generally of the

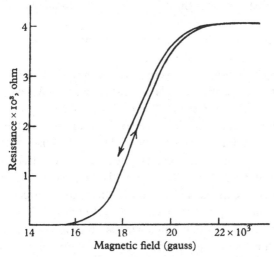

Fig. 14. Restoration of resistance of Pb-Bi eutectic by a magnetic field at 4·2° K. (de Haas and Voogd, 1930).

type shown in fig. 14, and differs strikingly from that of a pure superconductor. The most striking feature is the very high value of the field required to restore the first trace of resistance; this field varies widely for the different superconducting alloys (we have chosen for the diagram the example of the Pb-Bi eutectic which has one of the highest known critical fields), but is in general some thousands of gauss, instead of the few hundreds of gauss characteristic of most pure element superconductors. Another important feature is that the resistance is restored not in one jump but over a comparatively large range of fields (even if the magnetic field is parallel to the current in the alloy wire). On account of this spreading out of the transition, there is much less difference between the curves for a transverse and a parallel field than there is for a pure superconductor.

If after the resistance has been completely restored, the magnetic field is again reduced, there is usually a marked hysteresis (see fig. 14; the return curve is not drawn completed owing to lack of experimental data). This hysteresis is presumably connected with the hysteresis in the magnetic properties (i.e. absence of a Meissner effect) which we shall discuss below, but very few experimental data are available which could allow any conclusions to be drawn as to the more detailed nature of the connection; unfortunately, most experiments on alloys have been concerned with some particular aspect, and so there is no alloy for which all the different properties have been studied on one and the same specimen.

Before the magnetic properties were properly investigated, these apparently very high critical fields of alloys suggested several interesting possibilities. For instance, with the use of such an alloy, it seemed that it might be possible to realize the old idea of a super-conducting solenoid for fields of a few thousand gauss, since smaller fields, if applied externally, did not restore any trace of resistance. Also if H_c and dH_c/dT were as large as suggested by the resistance measurements, all the caloric effects to be discussed in Chapter III should appear on a scale several hundred times as large as for pure metals, so that an enormous jump in the specific heat should occur when superconductivity was destroyed in the alloy, and an enormous cooling effect if the destruction was in a magnetic field. In practice, however, it turned out that none of these possibilities could be realized, because the critical field suggested by the resistance measurements is characteristic of only a very small fraction of the bulk of the alloy. We shall now outline briefly the experimental evidence which points to this interpretation.

De Haas and Casimir-Jonker (1935), using the bismuth wire technique, showed that actually a magnetic field penetrated into an alloy long before it was large enough to restore the first trace of resistance, and that the penetration was very nearly complete at field strengths of the same order of magnitude as for pure elements. Similarly, Mendelssohn and Moore (1935a), and Rjabinin and Schubnikow (1935), measuring the B-H_e curve of a long rod of superconducting alloy, found that B ceased to be zero, and approached the value of H_e, at fields much lower than those required to restore the first trace of resistance (fig. 15).* This evidence at once suggests that the real critical field of most of an alloy specimen is of the same order of magnitude as for a pure superconductor, but that as the field is increased, a very small fraction of the whole bulk of the specimen, in virtue of a higher critical field, is able to remain superconducting, and so provide a path of zero resistance for the current used in the resistance measurements.

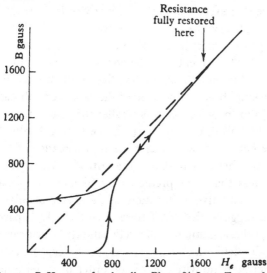

Fig. 15. B-H_e curve for the alloy Pb+2 % In at $T = 1\cdot95°$ K. (Schubnikow, Chotkewitsch, Schepelew and Rjabinin, 1936).

The failure of the Silsbee hypothesis (see § 1.4.2) for alloys also agrees qualitatively with the above interpretation; thus Keesom (1935) and Rjabinin and Schubnikow (1935), in attempting to make use of an alloy for the production of magnetic fields of a few thousand gauss, found that the resistance of an alloy wire was restored by currents much lower than would produce a field at the surface of the wire of the magnitude of the critical field as determined by resistance measurements. In the experiments of

* We have chosen a case where the penetration field is not so very much smaller than the field which restores resistance, in order to be able to show all the features in a single diagram. The region of penetration is however often very much greater than that shown in fig. 15.

Rjabinin and Schubnikow, it was shown that the field at the surface of the wire produced by the current which first caused a trace of resistance to return was independent of the diameter of the alloy wire, and was of the same order of magnitude (actually about 30% less for the particular alloy concerned) as the field which first penetrated the specimen if applied longitudinally. In terms of the above hypothesis, this probably means that the field of the current which begins to restore the resistance is equal to the lowest critical field of the alloy (supposing that on account of inhomogeneity there is a continuous distribution of critical fields). When this current is exceeded, superconductivity is destroyed in the whole wire in some way similar to that for a pure metal (§ 4.6), except that the process must be complicated by the inhomogeneity of the critical field. The current is unable to flow along the 'threads' of high critical field because, presumably, their radii are too small, i.e. roughly speaking, because the ratio of the radius of a thread of high critical field to the wire radius is smaller than the ratio of the critical field of the bulk of the wire to the highest critical field. The difficulty of the current flowing along the thin threads is further accentuated by the fact that probably the critical current in such thin threads is less than that predicted by Silsbee's hypothesis (see § 5.9). This qualitative explanation of the failure of Silsbee's hypothesis thus suggests that the failure is really only apparent, and that the hypothesis is still qualitatively valid if the critical field is taken not as the field which restores the resistance, but as that which first penetrates the wire, i.e. of the order of magnitude of the critical field of the bulk of the specimen.

Although the explanation given above of the differences between an alloy and a pure superconductor is plausible, it is only qualitative. So to avoid laying too much stress on this particular interpretation, it is convenient to recapitulate the facts from a purely experimental point of view. Thus we can say that in an alloy there are *three* critical fields: (1) the field H_1 produced at the surface of the wire by the current which just restores resistance, (2) the field H_2 which first begins to penetrate the specimen, and (3) the field H_3 which is required to restore resistance with only a weak current flowing. The temperature variation of these critical fields is shown in fig. 16 for a particular example.

For a pure metal these three critical fields are identical, but we see that, for an alloy, H_3 is much larger than H_2, while H_1 is slightly less (in this case by 30%) than H_2. In the interpretation of the failure of Silsbee's hypothesis, it was implied that there should be no difference between H_1 and H_2, and indeed it is possible that the field at which the penetration first occurs is really lower than that indicated by Rjabinin and Schubnikow, i.e. that the first penetration was too slight to be detected by their experimental arrangement. It is, however, possible that there is a real difference

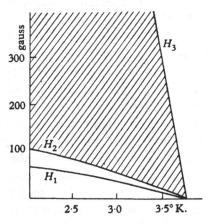

Fig. 16. Temperature variation of H_1, H_2 and H_3 for the alloy PbTl$_2$ (Rjabinin and Schubnikow, 1935).

between H_1 and H_2, and for the time being the question must be left open. The shaded region of fig. 16 is that in which penetration of the external field is becoming more and more complete, but usually the penetration is already very nearly complete soon after H_2 has been exceeded. The field H_3 and the form of the transition curve are very sensitive to the strength of the measuring current; for small measuring currents a considerable increase (of order 50%) of field above H_3 is necessary to restore the resistance completely (H_3 being only the field which restores the first trace of resistance), but for larger measuring currents the restoration is less spread out and H_3 is lower (Alekseyevsky, 1938b).

Fig. 15 shows another remarkable difference between the properties of an alloy and a pure superconductor, namely, that the

Meissner effect does not occur in alloys.* It will be seen, in fact, that there is a very marked hysteresis in the B-H_e curve when the magnetic field is reduced from a value at which it penetrates appreciably into the alloy, most of the flux which had penetrated into the specimen for this field value being frozen-in. The detailed form of the B-H_e curve depends on the particular alloy and also on the temperature; in contrast to the case of a pure super-conductor, the curves for different alloys, or for the same alloy at different temperatures (for similar shapes, of course), cannot be superimposed by a mere change of scale of B and H_e.†

This absence of a Meissner effect in alloys can also be roughly interpreted in terms of the hypothesis that in alloys there exist regions of abnormally high critical field, if the further hypothesis is made that these regions are multiply-connected. Mendelssohn (1935) has suggested, in fact, that the structure of an alloy resembles that of a sponge, whose meshes have a much higher critical field than that of the material they enclose. As the magnetic field is increased above the value corresponding to the critical field of the bulk of the alloy, it begins to penetrate into the specimen, but the penetration is hampered by the still superconducting meshes of the assumed sponge structure, which tend to keep the flux through the specimen constant. We have seen, however, in § 2.6, that a multiply-connected superconductor is unable to keep the flux constant in the material it encloses when the external field is too much increased, since it cannot carry more than a certain current; and, moreover, this limiting value of the external field is lower, the thinner the superconducting region. Thus, in spite of the high critical field of the meshes, on account of their thinness, they

* It may be noted that on this account the B-H_e curve of an alloy is closer to the theoretical curve for a perfect conductor (fig. 5b, Chapter II) than to that for a pure superconductor (fig. 6b, Chapter II).

† Some experiments on the magnetic properties of a tantalum specimen (which, presumably on account of impurities, behaved like an alloy) showed that the freezing-in was the more complete the lower the temperature was below the transition temperature, i.e. that the hysteresis was more marked at lower temperatures (Shoenberg, 1937a). The more extensive experiments of Schubnikow, Chotkewitsch, Schepelew and Rjabinin (1936), however, showed that this is not a general rule, since in some cases the hysteresis became *less* marked at lower temperatures. Possibly the difference is due to the fact that the tantalum specimen had a non-zero demagnetizing coefficient, while in the experiments of Schubnikow *et al.* long rods were always used.

are not able to prevent the flux through the specimen increasing with the external field, and eventually the penetration of the external field is nearly complete, although the meshes (or at least some of them) are still superconducting, and are able to pass a small current without any trace of resistance.

When the field is again reduced from some high value, the meshes are at first unable to freeze-in all the flux originally passing through the specimen, since this would require induced currents in the meshes larger than they could carry (again just as in the case of the ring). We have seen, in fact, that the meshes cannot carry a current higher than that which produces a field at the surface of the wire higher than the critical field of the bulk of the alloy. When the field, however, reaches the critical field of the bulk of the alloy (H_2), the superconducting meshes are able to grow thicker at the expense of the surrounding material, which was formerly in the normal state, but is now in a field less than its critical field. With increase of the thickness of the meshes they are able to carry more current, and thus to prevent most of the flux from leaving the alloy as the external field is reduced to zero; thus eventually the alloy specimen is left with a large fraction of the flux frozen-in which passed through it when the field had the value H_2, and, moreover, the bulk of its volume (through which this frozen-in flux passes) is left in the normal state. Further support for the general ideas discussed above is provided by the calorimetric evidence which will be described in § 3.4.

Although, on account of the inevitable inhomogeneity of the alloys of the type we have been considering, it is natural that they should have a *range* of critical fields (corresponding to the variety of compositions in different parts of the specimen), rather than a single critical field, there is as yet no complete explanation of why some small regions (the meshes of the sponge) should have such abnormally high critical fields as to remain superconducting up to several thousands of gauss. At first sight it might seem that these high critical fields were characteristic of some particular composition which occurred only in very small concentration in the specimens investigated. This idea, however, is unlikely to prove true, for it is unlikely that a very high critical field could ever be characteristic of a superconductor in bulk, since its entropy in the normal state (in a magnetic field, of course) would have to be enormously higher than in the superconducting state, and consequently the substance in the normal state would have to have a specific heat much greater than that of an ordinary metal (see Chapter III). No evidence has, in fact, been found for the existence of a substance in bulk with such high critical fields.

It is more probable, as we shall explain in more detail in § 4.6.2, that the survival of superconducting threads (i.e. the meshes of the sponge) at high fields, arises from a negative surface energy at normal superconducting interfaces in alloys (Gorter, 1935; H. London, 1935). For a pure metal this surface energy is positive, but it may become negative in regions of inhomogeneous strain such as are likely to occur abundantly in alloys, and we shall see in § 4.6.2 that in such regions of negative surface energy, superconductivity can persist in very thin threads or laminae well above the 'bulk' critical field of the material. On this view then, it is still the inhomogeneity of alloys which is basically responsible for their characteristic peculiarities. It is not, however, clear what scale of inhomogeneity is involved. The experiments of Stout and Guttman (1950), which show that even a disordered solid solution may, if sufficiently carefully prepared, behave nearly like a pure metal, suggest that perhaps disorder on an atomic scale is not sufficient to make the surface energy negative, while the theoretical considerations of Pippard (1950c, 1951a) (see § 6.3.5) suggest the opposite conclusion. More experimental evidence on the relation between the superconducting behaviour of alloys and their structure (phase constitution, atomic arrangement, etc.) is required to fill out the details of the interpretation.

In conclusion, we wish to mention one point of practical and historical, rather than theoretical interest. This is that the behaviour of alloys described above is in practice nearly always shown to a lesser extent by simple superconductors unless they are extraordinarily pure and free from mechanical strains. A great deal of the experimental work since the discovery of the Meissner effect has, in fact, been devoted to separating out the properties which we have, in the previous chapters, ascribed to pure superconductors, from the effects caused by minute traces of impurities, and this work has shown that actually very little impurity indeed (less than

0·001 % solid impurity in some cases) is necessary to produce some of the effects which we have described as characteristic of alloys. Thus for many of the element superconductors, it has not yet been possible to obtain a specimen which shows a complete Meissner effect (i.e. no hysteresis in its magnetization curve), in which the critical field as determined from resistance measurements is not higher than that determined from magnetic measurements, and in which the transition of a long cylinder does not take place over a short range of fields or temperatures, rather than at a single point (a slightly impure element never shows, however, the high critical fields characteristic of alloys). It is not yet clear which particular defects are most effective in causing a nominally pure superconductor to depart from its 'ideal' properties. Dissolved gases (not usually included in purity estimates), surface contaminations and internal strains may be relevant factors (see, for instance, Webber, 1947; Jackson and Preston-Thomas, 1950). The question of which metals approach ideal behaviour most closely will be discussed in § 2.11.

2.9. Problems of current distribution

In this section we shall discuss briefly a number of problems concerned with the distribution of current flow in various arrangements of superconductors in the absence of an applied magnetic field. Such problems are of little fundamental interest in themselves but are occasionally of practical importance in the design of experiments. The problem of destruction of superconductivity by currents fed in from outside will be dealt with in § 4.7, after a more thorough discussion of the intermediate state.

We have already pointed out (§ 2.3) that all currents flowing in superconductors are superficial, and are of density (per unit width) equal to $c/4\pi$ times the magnetic field at the surface (this magnetic field must be parallel to the surface). Thus the problem of finding the current distribution reduces to one of finding a distribution of magnetic field between the superconductors, consistent with the boundary conditions (e.g. total current given at entry and exit of the circuit), and such that the field is always tangential at all superconducting surfaces. This can best be dealt with by introducing the vector potential **A** (the scalar potential ϕ can be used only when there are no 'total' currents, as in §2.4), and the problem thus reduces to finding a suitable solution of the equation $\nabla^2\mathbf{A} = 0$.

Three examples will now be given which reduce to solutions of more familiar problems in electrodynamics, involving standard text-book methods:

2.9.1. *Current enters and leaves an infinite plane by two closed curves on the plane.* If we ignore the complications of the field due to the currents in the leads to the plane, the problem is essentially a two-dimensional one with the magnetic field parallel to the plane. If the z axis is taken as normal to the plane, $\partial/\partial z = 0$, and we have

$$H_x = -\partial A_z/\partial y, \quad H_y = \partial A_z/\partial x, \tag{2.23}$$

(this means that the lines of force are lines of constant A_z) and (2.11) becomes

$$g_x = \frac{c}{4\pi}\frac{\partial A_z}{\partial x}, \quad g_y = \frac{c}{4\pi}\frac{\partial A_z}{\partial y} \quad \text{or} \quad \mathbf{g} = \frac{c}{4\pi}\operatorname{grad} A_z, \tag{2.24}$$

which means that the lines of current flow are normal to the lines of constant A_z. The problem is thus one of finding a solution of

$$\frac{\partial^2 A_z}{\partial x^2} + \frac{\partial^2 A_z}{\partial y^2} = 0, \tag{2.25}$$

with A_z constant over the entry and exit curves (in order that **g** should be normal to them). This is identical with the problem of *ordinary* two-dimensional current flow, with A_z playing the role of ϕ, the electrostatic potential, and this in turn is identical with the two-dimensional electrostatic problem of finding the potential distribution between two infinite right metal cylinders with the given curves as cross-sections. A full discussion of methods of solution of this two-dimensional problem is given in standard text-books.

2.9.2. *Current flows along an infinite cylinder of arbitrary cross-section.* If the z axis is taken parallel to the cylinder axis, then $\partial/\partial z = 0$ and $H_z = 0$; the problem is again two-dimensional and, as before, (2.23) applies, so that the lines of force are again lines of constant A_z, and since H must be tangential at the surface of the wire it follows that we must find a solution of (2.25) such that A_z is constant over the surface. The problem is thus identical with the two-dimensional electrostatic problem of finding the charge distribution over an infinite metal cylinder, A_z again playing the role of the electrostatic potential, and the current g_z, which is clearly given by

$$g_z = \frac{c}{4\pi} \left| \operatorname{grad} A_z \right|, \qquad (2.26)$$

playing the role of the electric field (except, of course, that it is along, instead of normal to, the cylinder). This electrostatic problem is also discussed in standard text-books; the solution for a wire whose cross-section is an ellipse of minor axis $2a$ and major axis $2b$ may be recorded here, since it is useful in the discussion of a later problem (see § 5.9). It is

$$H = 4\pi g/c = 2i/(a^2 \cos^2 \theta + b^2 \sin^2 \theta)^{\frac{1}{2}}, \qquad (2.27)$$

where θ is the angle between the radius vector and the major axis of the ellipse. Thus the highest value of H is at the end of the major axis and the magnetic field there is $2i/a$.

2.9.3. *Distribution of current between parallel circuits.* If the various wires are separated by distances large compared with their diameters, the problem may be treated by introducing suitable coefficients of self- and mutual induction (von Laue, 1932); experiments by Sizoo (1926) and Justi and Zickner (1941) are in full accord with the theory. It may be mentioned that Meissner and Heidenreich (1936) have demonstrated the superficiality of the current by studying the detailed form of the field distribution round a system of superconducting wires (for a single wire the field distribution is of course independent of whether the current is superficial or not) (see also p. 30).

2.10. Miscellaneous magnetic effects

2.10.1. *Hall effect.* The only attempt to study the Hall effect in superconductors has been that of Onnes and Hof (1914), who showed that in the superconducting range there was no Hall e.m.f. greater than the limit of measurement.[*] Since, however, the Hall e.m.f. even in the normal state was less than the limits of measurement, this result is not of great significance. A simple argument due to Pippard (unpublished) suggests that even with sufficient sensitivity it would probably be impossible to detect a Hall effect by the conventional method, owing to compensating potential drops at

[*] For some interesting comments on this experiment, see Hall, 1933.

the surface which must be postulated to satisfy the law of conservation of energy. This argument is based on the idea that a current in a superconductor has the same character whether it is fed into the metal from outside, or flows in a closed path round the superconductor (as does the current associated with an applied magnetic field). Consider a sphere in a uniform applied magnetic field: if there is a Hall e.m.f. proportional to the vector product of current and field in the usual way, this will have a value proportional to the square of the field at the equator of the sphere, but will vanish at the poles. It would therefore appear to be possible to drive a current in a resistive circuit connecting the equator to a pole and thus draw energy continuously from the static magnetic field. This is inconsistent with the conservation of energy and so we must suppose that there is an opposing contact potential which varies with the field in such a way as to compensate any Hall e.m.f. exactly.

2.10.2. *The gyromagnetic effect* (see also § 6.2.6). The small angular momentum communicated to a superconductor when a magnetic field is applied has been measured by Kikoin and Goobar (1938, 1940) by a delicate resonance method. They found that the ratio of magnetic moment to angular momentum is just $-e/2mc$ (i.e. the Landé g-factor equals unity), which indicates that the diamagnetism of a superconductor is caused by ordinary electron currents and not, for instance, connected in any way with electron spin. This result is important in providing justification for the basic ideas of the Londons' phenomenological theory.

The inverse gyromagnetic effect—the 'Barnett effect', in contrast to the 'Einstein-de Haas effect' just described—is the production of a magnetic moment in a rapidly rotating superconductor. This effect will be discussed in § 6.2.6, where it will be shown that, as in the ordinary Barnett effect for a substance with $g = 1$, the expected magnetic moment should be just that produced by a magnetic field

$$H = 2mc\omega/e,$$

where ω is the angular velocity of the superconductor. It can be seen that this field is very small for any practicable value of ω (e.g. $H \sim 10^{-4}$ gauss for $\omega = 10^3$ sec.$^{-1}$), and so far its existence has not been demonstrated experimentally, though attempts are in progress (Squire and Love, 1949).

2.10.3. *Other magnetic properties of moving superconductors.*
Experiments by Condon and Maxwell (1949) have shown that the
currents causing the magnetic properties of superconductors are
rigidly linked to the magnetic field rather than the specimen.
A sphere was suspended by a torsion wire in a uniform horizontal
field, and it was shown that the period of torsional oscillations was
independent of the field strength, provided the specimen was
wholly superconducting. Had there been any twisting of the
system of currents in the specimen as it rotated, a couple would
have been exerted by the stationary field which would have modified
the period.* This result is not very surprising, since the period of
oscillation was quite low and relaxation effects in which the
equilibrium distribution of currents did not follow instantaneously
the changes of the field direction relative to the specimen would be
expected only for very high frequencies of oscillation—of the order
of those which produce resistive effects (see § 1.4.6).

Another type of experiment, which, however, really does little
more than confirm the Meissner effect in a complicated way, is to
rotate a sphere in a magnetic field parallel to the axis of rotation,
and to study the e.m.f. between rubbing contacts at the equator
and a pole. In the normal state, the sphere acts rather like
a Faraday disk, and an easily calculable e.m.f. should be observed
between the contacts due to the cutting of the magnetic lines of
force. As soon, however, as the sphere becomes superconducting,
this e.m.f. should disappear, since practically no lines of force are
then left in the sphere to be cut. This experiment was first tried
by Houston and Squire (1949a), but complicated results were
obtained owing to technical difficulties (friction at the bearings
causing rise of temperature); repetitions by Wexler and Corak
(1949, 1950a) and Houston and Squire (1949b) gave results closely
in agreement with prediction.

2.11. Experimental methods

The first problem in any experiment on superconductivity,
which is concerned with establishing general results rather than the
specific properties of a particular metal, is to prepare a suitable

* Much more complicated effects occur, however, if the specimen has a
frozen-in moment, if its shape is not accurately spherical, or if it is in the
intermediate state (see, for instance, Fritz, Gonzalez and Johnston, 1949, 1950).

specimen. This must be as free as possible from any defects which may introduce secondary features and thus confuse observation of what we have described frequently as 'ideal' behaviour. As explained in § 2.8, experience has shown that chemical purity (including freedom from dissolved gases) and freedom from internal stresses are essential, and these qualities are in practice to be found in only a few metals, mostly the 'soft' superconductors. In some experiments (for instance, on high-frequency effects (§ 1.4.6), and size effects (Chapter v)), surface conditions are very important, and this usually makes easily oxidizable metals, such as lead and thallium, unsuitable. Sometimes it is found that machined surfaces must be avoided in favour either of cast or electrolytically polished ones (see, for instance, § 5.3.1). Practical considerations often weigh heavily against metals with transition temperatures which are either very low (say below 1·5° K.) or above 4·2° K., since the control, maintenance and measurement of such temperatures involve the use of special techniques which complicate the experiment. These considerations leave only mercury, tin and indium, and of these, pure indium is very expensive, while mercury is awkward to handle since it is liquid at ordinary temperatures, so that we are finally left with tin as the only generally suitable superconductor. Fortunately, spectroscopically pure tin is readily available, easily machined, cast, or grown into single crystals, and can be electrolytically polished.

The various methods which have been used for studying magnetic properties will now be briefly indicated; there is no need to go into any detail, since they are for the most part merely modifications of standard magnetic methods.

2.11.1. In the original experiments of Meissner and Ochsenfeld (1933) the field at various places round the specimen was measured by means of search coils which could be turned over. This method (also used by Tarr and Wilhelm (1934) in experiments which confirmed the Meissner effect) has the disadvantage that it gives only the average field over the area of the search coil, and it involves somewhat complicated devices for setting and rotating the coil, but it is useful in special cases—for instance, if a large specimen can be used.

2.11.2. The resistance of a bismuth wire depends strongly on magnetic field, particularly at low temperatures, and this forms the basis of a convenient method of studying field distributions which was first applied to problems of superconductivity by de Haas and Casimir-Jonker (1934). The already mentioned experiments of Mendelssohn and Babbitt (1935) and de Haas and Guinau (1936) also made use of such bismuth probes, while Shalnikov (1945) and Meshkovsky and Shalnikov (1947a, b), in their studies of the structure of the intermediate state (see § 4.3), have refined this method into a microprobe technique capable of resolving fluctuations of field on a very fine scale.

2.11.3. Rjabinin and Schubnikow (1934) measured the total induction of the specimen by removing it suddenly from inside a long coil, and observing the ballistic throw of a galvanometer connected in series. They also sometimes used an almost identical method, in which the throw was observed for a sudden change of field, the specimen remaining in the coil. Their experiments were made with long cylinders parallel to the field, and are of interest as confirming the Meissner effect in its application to the magnetization curve of a long cylinder (i.e. confirming the behaviour described in § 2.3). The same type of method has been used by Mendelssohn and his collaborators in various investigations.*

More recently this method has been developed to give the great sensitivity necessary for studying the magnetic behaviour of specimens of small size (Shoenberg, 1947; Désirant and Shoenberg, 1948a, b). Two coils in opposition are used so that the pair is insensitive to changes of field in the surrounding solenoid, and by putting the coils in liquid helium the number of turns can be greatly increased without exceeding the critical damping resistance of the galvanometer. If now the specimen is pulled smartly from the centre of one coil to the centre of the other, while remaining all the time in a uniform field, the deflexion of the galvanometer is proportional to the magnetic moment. The sensitivity is further increased by photoelectric amplification of the galvanometer deflexion, and it is actually used as a null method, the magnetic

* See, for instance, Mendelssohn and Moore, 1935a, Keeley and Mendelssohn, 1936, Mendelssohn, 1936, Mendelssohn and Pontius, 1936a,b, Daunt and Mendelssohn, 1937.

moment of the specimen being compensated by adjusting a current in a small auxiliary coil which surrounds the specimen and moves with it.

2.11.4. In the already mentioned experiments on spheres, rings and short cylinders, Shoenberg (1935, 1936, 1937 a) measured the force on the specimen in a slightly inhomogeneous field; the force divided by the field gradient then gives the magnetic moment. This is, of course, nothing else than the well-known Faraday method of determining susceptibilities, and is reasonably simple to apply owing to the very strong diamagnetism of superconductors, so that the balance used for measuring the force has not to be particularly sensitive.

The induction method described above is, however, more accurate, avoids the difficulties associated with an inhomogeneous field, and is simple to use. The application of the force method to superconductors is, in fact, now of little more than historic interest, as having demonstrated that a superconductor does really behave like a very strong diamagnetic.

2.11.5. A non-spherical specimen in a homogeneous magnetic field in general experiences a couple due to anisotropy of its demagnetization coefficient. Thus a magnetic field exerts a strong couple trying to twist a superconducting plate into the field direction. This has been used by Alekseyevsky (1941 b) to study the magnetic properties of thin superconducting films (see § 5.2.3).

2.11.6. The effective alternating field susceptibility of a specimen in various circumstances can be studied by measuring the change of either the self-inductance of a coil (Shoenberg, 1937 b) or the mutual inductance of a pair of coils (Daunt, 1937) surrounding the specimen. In the experiments of the author the main aim was to study the differential susceptibility (dI/dH) of a sphere in the intermediate state, but the results were complicated by the appearance of energy losses due to non-zero resistance in the intermediate state. In a special experiment with a sphere made up of thin laminations the effect of the eddy currents was greatly reduced, and the specimen did indeed behave approximately as if the magnetic properties in the intermediate state were given by (2.17) with dI/dH changing its sign and roughly doubling its magnitude on entry into the intermediate state. We shall return to some other features of these experiments in § 4.4.

In the purely superconducting state these experiments confirmed that the permeability was zero; more recently the same principle applied in a refined form has been used to measure the temperature dependence of the slight departure from zero permeability due to the temperature dependence of penetration depth (Casimir, 1940; Laurmann and Shoenberg, 1947, 1949). This application will be discussed in § 5.3.1.

In conclusion, we may mention that all of these magnetic methods, but particularly those of §§ 2.11.3 and 2.11.6, are often much more convenient than the electrical method (i.e. measurement of resistance) for testing for superconductivity, for determining critical fields, and for examining how far a specimen behaves like an 'ideal' superconductor. An obvious advantage of the magnetic methods is that no leads have to be attached to the specimen, and for some purposes, moreover, the shape of the specimen may be left quite arbitrary, so that the material to be tested can be used in the form in which it is available, without any mechanical treatment. Another advantage is that the magnetic methods generally give some idea of the volume properties of the specimen, while the electrical method gives effectively the resistance only of the path of least resistance; thus we have seen (§ 2.8) that superconducting alloys may have zero resistance in very high magnetic fields, while the magnetic methods show that actually only a very small fraction of the volume is still superconducting. The use of a magnetic method to search for new superconductors was first applied by Kürti and Simon (1935) in the range of very low temperatures opened up by the paramagnetic salt demagnetization technique and three new superconductors were found in this way.

CHAPTER III

THERMODYNAMIC AND OTHER THERMAL PROPERTIES

3.1. Differences of thermodynamic functions

The idea of applying thermodynamics to the transition between the superconducting and normal states was originally suggested by Keesom (1924) and then by Rutgers (see Ehrenfest, 1933; and Rutgers, 1934, 1936), and developed in greater detail by Gorter (1933) (see also Gorter and Casimir, 1934a). At that time it was still thought that the transition was essentially irreversible in a magnetic field, since a superconductor was believed to be merely a perfect conductor (in the sense explained in Chapter II), so that the surface currents associated with the field would die away with the production of Joule heat when superconductivity was destroyed by the field. This supposed irreversibility made the validity of Gorter's treatment very doubtful, and it seemed surprising that his results should agree so well with experiment. The reason for this agreement appeared shortly afterwards, with the discovery of the Meissner effect, which showed that the disappearance of the super-conducting surface currents in a pure metal is, in fact, not associated with any irreversible energy changes (somewhat analogously to the diminution of the magnetization of a ferromagnetic when its temperature is raised), and that the basic assumption of reversibility in Gorter's treatment is indeed valid.

In this chapter we shall deal only with macroscopic effects, in the sense explained in §2.1; the modifications brought about by consideration of surface effects (which are of importance in the intermediate state) will be mentioned in Chapter IV, while those due to penetration effects, which are of importance for small-sized specimens, will be discussed in Chapter V. The thermodynamics of the superconducting transition in a magnetic field is perfectly analogous to that of any other phase transition, and all the results may be obtained in the simplest possible way by equating the free energies of the two phases. To obtain the free energies, the simplest approach is to treat the metal in the superconducting state as a magnetic substance having the properties described in Chapter II, and in the normal state as a non-magnetic substance. If the Gibbs free energy, $U - TS + pV - H_e IV$, of a specimen of volume V in the absence of a magnetic field is G_s in the superconducting state, then in an applied field H_e it is*

$$G_s(H_e) = G_s - V \int_0^{H_e} I \, dH_e.$$

If the Gibbs free energy in the normal state is G_n, we can suppose that this is independent of H_e, since the normal state is practically non-magnetic (the actual feeble diamagnetism or paramagnetism is negligible compared with the strong diamagnetism of the super-conducting state). As we saw in § 2.5, it is only for a body of zero demagnetizing coefficient, e.g. a long rod parallel to the field, that the destruction of superconductivity occurs sharply at a well-defined field; and for this case we can see that the transition must therefore occur at a field H_c such that

$$G_s(H_c) = G_n,$$

or since $I = -\dfrac{1}{4\pi} H_e$, when

$$G_n - G_s = V H_c^2 / 8\pi. \tag{3.1}$$

Although this relation has been derived by considering a body of zero demagnetizing coefficient it clearly cannot depend on the shape of the body, since G_s is defined in the absence of a field. For a body of non-zero demagnetizing coefficient, the transition is spread over a range of applied fields, corresponding to the intermediate state, in which, as mentioned in § 2.5, the specimen is split up into super-conducting and normal regions. This mixture is quite analogous to the mixture of vapour and liquid, for instance, which occurs at a certain pressure (which like H_c is a function of temperature) for which the free energies of vapour and liquid are equal. For the liquid-vapour transition, the relative amounts of the two phases are determined by the total volume of the system, while in the inter-mediate state of a superconductor, the relative amounts of the two phases are given by the total induction of the specimen, i.e. by B.

It was shown in § 2.5 that the area of the magnetization curve is always $H_c^2 / 8\pi$ whatever the shape of the body, and we shall now

---

* For simplicity it is assumed here that I is uniform, i.e. that we are dealing with an ellipsoid; the more general expression is given at the top of p. 58.

show that this is a consequence of (3.1). The free energy of a body of *any* shape is

$$G(H_e) = G_s - \int dV \int_0^{H_e} I \, dH_e,$$

even if H_e is big enough to destroy part or all of the superconductivity. In particular, if $H_e > H_c$, superconductivity is completely destroyed and

$$G(H_e) = G_n,$$

so we have

$$G_n = G_s - \int dV \int_0^{H_e} I \, dH_e$$

(since for $H_e > H_c$, $I = 0$), and using (3.1), we see at once that $-\dfrac{1}{V} \int dV \int_0^{H_c} I \, dH_e$, the area of the magnetization curve, is always $H_c^2/8\pi$, whatever the specimen shape.* The thermodynamic properties of the intermediate state will be discussed further in §3.3.

From (3.1) we can deduce the differences of all the thermodynamic quantities for the two phases; thus differentiating with respect to temperature, we obtain the difference between the entropies of the normal and superconducting phases, since $S = -dG/dT$. We find

$$S_n - S_s = -\frac{VH_c}{4\pi}\frac{dH_c}{dT}. \tag{3.2}$$

Since dH_c/dT is always negative, we see that the entropy of the superconducting phase is less than (or equal to, when $H_c = 0$) that of the normal phase, or in other words superconductivity corresponds to a more ordered state of the metal. The entropy difference vanishes at the normal transition temperature T_c, i.e. in the absence of a magnetic field, and by Nernst's theorem it must vanish also at absolute zero. The last circumstance means that dH_c/dT must vanish at absolute zero, and this is in agreement with the experimental evidence as far as it goes (the measurements have not, however, been carried down to sufficiently low temperatures to make the confirmation entirely certain). Since the entropy difference vanishes both at $T = T_c$ and at $T = 0$, it must pass somewhere through a maximum; this is illustrated by the curves of fig. 17.

* This argument assumes that only reversible processes are involved in the transition; if the transition for increasing fields is irreversible the area will be greater than $H_c^2/8\pi$, and if it is irreversible for decreasing fields, the area will be less than $H_c^2/8\pi$.

From equation (3.2) we see also that heat is absorbed when superconductivity is destroyed isothermally in a magnetic field, and if the change is made adiabatically, the temperature will drop; we shall discuss the magnitude of this cooling effect later. The heat absorbed in a transition at temperature T from the superconducting to the normal state is $Q = T(S_n - S_s)$, or

$$Q = - VT \frac{H_c}{4\pi} \frac{dH_c}{dT}. \tag{3.3}*$$

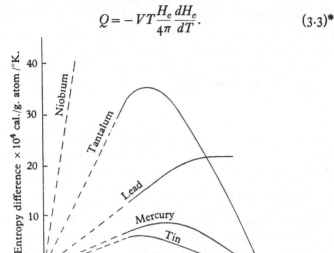

Fig. 17. Entropy difference as a function of temperature for various pure superconductors (Daunt and Mendelssohn, 1937). Note that the entropy differences are of order $10^{-3} R$, where R is the gas constant.

This result is quite analogous to the well-known Clausius-Clapeyron formula, the variables p and V being replaced by H_c and $- I_0 V$ respectively. We see that this heat vanishes at the normal transition temperature T_c, since dH_c/dT remains finite at T_c, so that there is no heat of transition in the absence of a magnetic field. This has already been mentioned in § 1.2.4 as an experimental fact.

Differentiating (3.2) with respect to T, we find the difference of heat capacity (i.e. the specific heat if V is the specific volume, or the atomic heat if V is atomic volume) of the normal and superconducting phases:

$$\Delta C = C_s - C_n = \frac{VT}{4\pi} H_c \frac{d^2 H_c}{dT^2} + \frac{VT}{4\pi} \left(\frac{dH_c}{dT}\right)^2. \tag{3.4}$$

* This result was first obtained by Keesom (1924).

In particular, in the absence of a magnetic field (i.e. for $T = T_c$), we have

$$\Delta C = \frac{VT_c}{4\pi} \left(\frac{dH_c}{dT}\right)^2_{T=T_c},$$ (3.5)

which is known as Rutgers's formula (Ehrenfest, 1933), and shows that if the metal is cooled or warmed in the absence of a magnetic field, there will be a discontinuity in its specific heat in passing the transition temperature. In the absence of a magnetic field, the specific heat of the superconducting phase is evidently greater than that of normal phase, but at lower temperatures (i.e. when superconductivity is destroyed in a magnetic field) the sign of $C_s - C_n$ must change, corresponding to the fact that $S_n - S_s$ passes through a maximum. It is important to notice that the superconducting transition in a magnetic field is an ordinary phase transition of the first kind, but in the absence of a field it degenerates into a transition of the second kind in the sense first formulated by Ehrenfest (1933).*

3.2. Experimental results

The jump in the specific heat was first found experimentally by Keesom and Kok (1932) for tin (actually a short time before the theory had been developed), and as can be seen from fig. 18 a it is exceedingly sharp. Indeed, the superconducting transition in the absence of a magnetic field provides probably the best example of an 'ideal' second-order transition that has yet been observed; in most other examples of second-order transitions the jump is more spread out. The magnitude of the jump at T_c for tin and nearly all the other metals studied agrees very well with Rutgers's formula, as can be seen from Table I, and as can be seen from fig. 18 b the values of ΔC at lower temperatures agree well with (3.4).† The equation (3.3) for the latent heat of transition in a magnetic field has also been confirmed experimentally for thallium (Keesom and Kok, 1934 b, c) and for tin (Keesom and van Laer, 1937). The satisfactory agreement

* For a full discussion of the classification of phase changes see Landau and Lifshitz, 1938.

† In fig. 18 a the earlier data of Keesom and Kok (1932) have been chosen for illustrating the specific-heat jump, since the data of Keesom and van Laer (1938) shown in fig. 18 b, though more detailed at lower temperatures do not cover the region of the transition temperature in so much detail. For a discussion of the agreement of (3.4) with the experimental data see also Keesom and van Laer (1938), Daunt, Horseman and Mendelssohn (1939) and Misener (1940).

THERMODYNAMIC AND THERMAL PROPERTIES

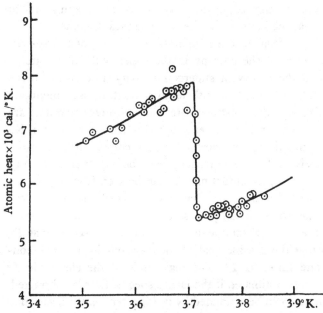

Fig. 18a. Temperature variation of the atomic heat of tin
(Keesom and Kok, 1932).

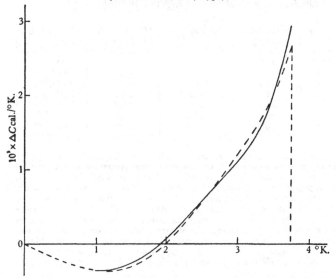

Fig. 18b. Temperature variation of ΔC for tin. ---- calculated from (3.4) using empirical H_c-T relation given in Appendix I, Table IX. —— smoothed experimental values (Keesom and van Laer, 1938).

between theory and experiment confirms the assumption of reversibility at the basis of the thermodynamical treatment.

The detailed form of the temperature variation of the specific heat, and hence of the entropy in the superconducting state, is important in the theory of superconductivity, but owing to the smallness of the specific heat, sufficiently accurate measurements are difficult. Perhaps the most accurate and comprehensive data are those of Keesom and van Laer (1938) for tin, down to $1.2°$ K., which show that the specific heat in the superconducting state varies approximately as T^3 over practically the whole temperature range in which tin is superconducting. As can be seen from fig. 19 the departures of C/T^3 from constancy are only slightly greater than the scatter of the experimental points. Since at sufficiently low temperatures the specific heat due to lattice vibrations also varies as T^3, it follows that the specific heat of the electrons in the superconducting state varies as T^3. The magnitude of the electronic T^3 variation can be estimated if the lattice specific heat is subtracted, and this in turn can be estimated by fitting a temperature variation

TABLE I. *Verification of Rutgers's formula* (3.5)

Metal	T_c ° K.	V (cm.^{3}/g. atom)	$(dH_c/dT)_{T=T_c}$	$\Delta C \times 10^3$ calc. (cal./° K.)	$\Delta C \times 10^3$ obs. (cal./° K.)
Pb	7·22	17·8	200	10	12·6[1]
Ta	4·40	10·9	320	9·4	8·2[2], 9[7]
La	4·37	22·4	> 1000	> 190	13·9[8]
Sn	3·73	16·1	151	2·61	2·4[8], 2·9[5]
In	3·37	15·2	146	2·08	2·3[1]
Tl	2·38	16·8	139	1·47	1·48[4]
Al	1·20	9·9	177	0·71	> 0·46[6]

Notes. The data for T_c and dH_c/dT are taken from the Appendix (Tables VIII and IX and fig. 77); the values of V are corrected approximately for the thermal contraction between room temperature and T_c. The values of ΔC obs. are as given in the original papers and represent all the data available at the time of writing. In most cases the scatter of the experimental points is considerable and an error of 5 or 10 % in ΔC obs. is quite possible. For Al it is possible that the transition was incomplete at the lowest temperature measured, so the value of ΔC obs. is probably only a lower limit; for La the disagreement between the observed and calculated values suggests that the critical field measured was much higher than the true equilibrium field (see Appendix I, notes to fig. 77 and Table IX, p. 227)—the value of $(dH_c/dT)_{T=T_c}$ calculated from ΔC obs. is 270 gauss/° K.

References. (1) Clement and Quinnell (1950, 1952), (2) Keesom and Désirant (1941), (3) Keesom and Kok (1932), (4) Keesom and Kok (1934a), (5) Keesom and van Laer (1938), (6) Kok and Keesom (1937), (7) Mendelssohn (1941), (8) Parkinson, Simon and Spedding (1951).

of the form $aT + bT^3$ (see (3.6)) to the data for the specific heat of the metal in the normal state. In this way Keesom and van Laer found a value $0.73 \times 10^{-4} T^3$ cal./°K./g. atom for the lattice specific heat, which if subtracted from the total specific heat of the metal in the superconducting state (about $1.7 \times 10^{-4} T^3$ cal./°K./g. atom; as can be seen from fig. 19), leaves about $1.0 \times 10^{-4} T^3$ cal./°K./g. atom as the electronic contribution. Measurements on lanthanum (Parkinson, Simon and Spedding, 1951) also suggest a T^3 variation in the superconducting state, and so do measurements on thallium (Keesom and Kok, 1934a), and on indium (Clement and Quinnell,

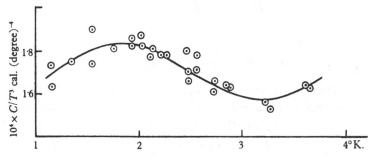

Fig. 19. Plot of C/T^3 against T (C is the atomic heat of tin, taken from data of Keesom and van Laer (1938)).

1950) though less conclusively owing to the more limited temperature range covered. It is therefore probable that, at least to a fair approximation, a T^3 variation is a general property of superconductivity.

Kok (1934) first pointed out that some interesting consequences may be deduced thermodynamically if the specific heat of the metal in the superconducting state is assumed to vary as T^3, and the specific heat of the normal phase (produced by applying a sufficiently large magnetic field) is assumed to vary as $aT + bT^3$. The latter assumption is, of course, reasonable if it can be assumed that the lattice specific heat does indeed vary as T^3,* and that the electronic specific heat in the normal state varies as T; it is also reasonably consistent with the experimental data. It must not be forgotten, however, that the theoretical derivation of the linear term is based on an electron

---

* For a detailed discussion of the conditions in which the lattice specific heat varies as T^3 see Mott and Jones, 1936, Chapter 1.

gas model, which is only a crude description of a metal, and cannot account for superconductivity.

Let us assume then that

$$\frac{C_s}{V} = BT^3, \quad \frac{C_n}{V} = aT + bT^3. \tag{3.6}$$

Integration, bearing in mind Nernst's theorem, gives immediately

$$\frac{1}{8\pi} \frac{d(H_c^2)}{dT} = \frac{S_s - S_n}{V} = \tfrac{1}{3}(B-b)T^3 - aT.$$

Now at $T = T_c$ there is no heat of transition, so we must have

$$\tfrac{1}{3}(B-b)T_c^2 = a, \tag{3.6a}$$

or

$$\frac{d(H_c^2)}{dT} = 8\pi a\left(\frac{T^3}{T_c^2} - T\right).$$

Integrating once more,

$$H_c^2 = 4\pi a\left(\frac{1}{2}\frac{T^4}{T_c^2} - T^2\right) + H_0^2,$$

where H_0 is the critical field for $T=0$, and putting in the condition that $H_c = 0$ for $T = T_c$, we find

$$a = H_0^2/2\pi T_c^2, \tag{3.7}$$

and

$$H_c = H_0(1 - (T/T_c)^2). \tag{3.8}$$

We see therefore that on the basis of the assumptions (3.6), not only is it possible to predict the parabolic form (3.8), which as mentioned in § 1.4.1 is in fair agreement with experiment, but the value of the coefficient a in the linear term of the specific-heat variation can be deduced from purely superconducting data. The experimental observation (see § 1.4.1) that H_0/T_c is not very different as between different superconductors in the 'soft' group means simply that a does not vary much.

It is important to notice that although (3.7) and (3.8) can be deduced from (3.6) by successive integration, the reverse procedure is not possible, for evidently since it is only the *difference* of C_n and C_s which has been integrated, it is only the difference which can be deduced if (3.8) is successively differentiated. Indeed, all that can be deduced if (3.8) is substituted in the general result (3.4) is

$$\frac{C_s - C_n}{V} = \frac{H_0^2 T}{2\pi T_c^2}\left(3\left(\frac{T}{T_c}\right)^2 - 1\right), \tag{3.9}$$

from which (3.7) follows only if it is *assumed* that C_s is proportional to T^3, i.e. that the linear term is *entirely* due to C_n.

Daunt and Mendelssohn (1937) have obtained estimates of a for various metals by a treatment of the critical field data which is essentially the same as the application of Kok's formula (3.7), but involves less assumptions. Using (3.2) they constructed curves of the temperature variation of entropy difference $S_n - S_s$, from specially made accurate measurements of the H_c-T relation for the metals concerned. These are the curves illustrated in fig. 17, and it can be seen that for sufficiently low temperatures the entropy difference does appear to vary linearly with temperature, though in fact it was not possible to take the measurements to quite low enough temperatures to make this conclusion certain. If then it is assumed that the entropy of the superconducting phase disappears much more rapidly than linearly, so that it is negligible at the temperatures concerned, the slope of the linear portion immediately gives a. This procedure assumes slightly less than that of Kok, in so far as Kok assumes that the entropy of the superconducting phase varies specifically as T^3, while Daunt and Mendelssohn require only that it should be negligible at sufficiently low temperatures. The value of a deduced is, of course, for a parabolic H_c-T curve, identical with that given by Kok's formula (3.7), but Daunt and Mendelssohn's procedure requires only that the relation is parabolic near $T = 0$ (this is implied by the linearity of $S_n - S_s$ with T for low temperatures).

3.3. Thermodynamics of the intermediate state

The formulae (3.3) and (3.4) give the heat of transition and specific heat difference for a *complete* transition between the superconducting and normal phases, but, as we have already pointed out, if the specimen has a non-zero demagnetizing coefficient, this transition is actually spread out over a finite range of applied fields if at constant temperature (the field at the superconducting boundaries is, however, constant and equal to H_c, throughout the transition), or over a finite range of temperatures if at constant external field (in this case, the field at the superconducting boundaries is also equal to H_c throughout the transition, but this H_c varies with the temperature—the analogy is that of a gas liquefying at varying pressure). Thus if the transition is made at constant temperature with an ellipsoid of demagnetizing coefficient $4\pi n$, the transition will take place gradually as the external field is increased from $(1-n)H_c$ to H_c, and correspondingly, the heat Q of (3.3) will be absorbed gradually. In a similar way, (3.4) gives only the difference of the specific heats before the transition began and after it is complete, but says nothing about how the specific heat will vary during the transition. In order to see how the specific heat varies during the transition process, we shall consider how much heat has to be supplied to the ellipsoid to raise its temperature by dT, the external field H_e being kept constant at a value in the intermediate range. In § 2.5 it was explained that in the intermediate state we have a mixture of the normal and superconducting phases, and it is easily seen that the proportions must be x and $1-x$, where

$$x = B/H_c. \tag{3.10}$$

Now owing to the rise of temperature, H_c will decrease slightly and so x will increase slightly, i.e. the proportion of the normal phase will be slightly increased. This transformation of a fraction dx of the specimen from the superconducting to the normal state will, however, require heat $Q\,dx$, so that more heat will have to be supplied to produce the temperature rise than would be required in either of the pure phases alone. In other words, the heat of transition, instead of appearing directly, shows up as an anomaly— obviously an increase—in the specific heat.[*] If it were not for the heat absorbed in the transition, the specific heat in the intermediate state would evidently be $xC_n + (1-x)C_s$, so the specific heat that should actually be observed is

$$C = xC_n + (1-x)C_s + Q\,dx/dT. \tag{3.11}$$

Substituting the value of x from (3.10), the value of B from (2.17), i.e. $B = H_c - (H_c - H_e)/n$, and the value of Q from (3.3), we find for the specific heat of the metal in the intermediate state

$$C = C_n\left(1 - \frac{1}{n}\left(1 - \frac{H_e}{H_c}\right)\right) + C_s\left(\frac{1}{n}\left(1 - \frac{H_e}{H_c}\right)\right) + \frac{VT\,H_e}{4\pi n\,H_c}\left(\frac{dH_c}{dT}\right)^2.$$

$$\tag{3.12}\dagger$$

---

[*] There is some analogy between these considerations and those used in deducing the specific heat of a saturated vapour.

† This result was first obtained by Peierls (1936).

We see, therefore, that at $H_e = (1-n)H_c$ the specific heat will jump upwards from the value C_s to $C_s + \dfrac{VT}{4\pi n}(1-n)\left(\dfrac{dH_c}{dT}\right)^2$; it will then increase linearly with H_e until the external field reaches H_c, when it will jump downwards from the value $C_n + \dfrac{VT}{4\pi n}\left(\dfrac{dH_c}{dT}\right)^2$ to C_n. If, instead of measuring the specific heat, we merely measure the absorption of heat as superconductivity is destroyed by an increasing magnetic field, we shall find that for an increase dH_e of the external field, heat $Q\,dx$ is absorbed, where dx is the fraction of the metal which has changed from the superconducting to the normal phase by the increase of field; it is easy to see that this heat is equal to $\dfrac{VT}{4\pi n}\left(\dfrac{dH_c}{dT}\right)dH_e$, i.e. that the heat is absorbed uniformly (the quantity for change dH_e not depending on the value of H_e). The variation of the specific heat with temperature in a constant external field is also given by equation (3.12), where now T (and therefore H_c) is the variable instead of H_e; there will again be two discontinuities at those temperatures for which the constant external field corresponds to $(1-n)H_c$ and H_c respectively. It will be noticed that (3.12) predicts infinite discontinuities for the case $n = 0$ (infinite cylinder or thin disk parallel to the field), but this need not alarm us, since these two discontinuities are infinitely close together, and simply indicate that the whole heat of transition is absorbed at a single field strength or temperature. In this limiting case the transition is quite sharp, and the formulae (3.3) and (3.4) are entirely adequate, so there is, in fact, no point in applying the considerations for a smeared-out transition.

These predictions have not yet been experimentally verified for an ellipsoid, but Keesom and Kok (1934b, c) have made measurements of the specific heat of a thallium specimen of irregular shape in a magnetic field which provide a qualitative confirmation of the theory.* As we have already pointed out, a specimen of irregular shape has an irregular field distribution round it, and so, unlike an ellipsoid, is unable to enter the intermediate state uniformly at any

* Owing to a slightly incomplete Meissner effect, more complicated results were obtained if the magnetic field was applied before the specimen had become superconducting. The results quoted apply to a field put on after the specimen had been cooled in zero field.

definite value of the external field. The result of this is in general to make the penetration of the external field into the specimen more gradual, so that the magnetization curve, for instance, has no sharp corner; consequently, the discontinuities of the specific-heat curve are smoothed out, and the transition heat shows up merely as a smooth hump in the specific-heat curve. In fig. 20 we reproduce the experimental curve of Keesom and Kok, together with the curve (shown by the broken line) that we should expect for an ellipsoid comparable with their block (we have assumed $n = o \cdot 1$).

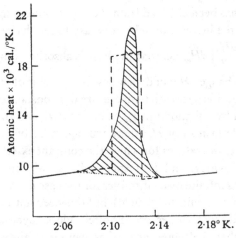

Fig. 20. Temperature variation of the atomic heat of a thallium block in a magnetic field of 33·6 gauss (Keesom and Kok, 1934c).

If the temperature range in which the hump occurs is relatively small (as in fig. 20), it is possible to deduce the total latent heat of the transition from the specific-heat curve. Thus if (3.11) is integrated over this small temperature range, C_n and C_s may be treated as constant, and if $(C_n - C_s)$ is small compared with C_s and C_n (as in fig. 20) no great error is made by assuming that x varies linearly with the temperature. If this assumption is made, it is easily shown from (3.11) that the transition heat Q will be given by the area of the shaded region in fig. 20 (more precisely, this gives an average of Q over the temperature range concerned). In this way Keesom and Kok obtained values of Q in good agreement with (3.3), thus again confirming the reversibility of the superconducting transition. Alternatively, knowing the value of Q, the value of x at any stage of the transition can be obtained as the ratio of the shaded area, up to the temperature concerned, to Q.* In this way the curve of fig. 21, showing the variation of x with T, was obtained (the curve for an ellipsoid is shown by the broken line, which is approximately straight on account of the small range of temperatures); we may point out again that it is because of the rounding off† of the corners of this curve

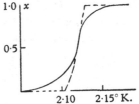

Fig. 21. Temperature variation of x (Keesom and Kok, 1934c).

that the discontinuities in the specific-heat curve are smoothed out.

As another illustration of the application of thermodynamics to the intermediate state, we shall discuss the magneto-caloric effect, i.e. the change of temperature for adiabatic change of H_e. Fig. 22 shows the entropy diagram for tin, calculated from the data of

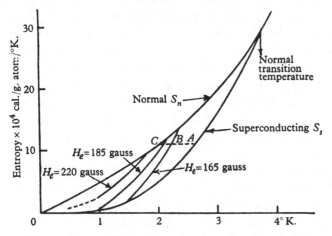

Fig. 22. Entropy diagram for a tin sphere (based on data of Keesom and van Laer, 1938).

* This is, of course, only a first approximation; a better approximation is obtained if the first approximation is used to correct the assumption (in the first two terms of (3.11)) that x varies linearly with T.

† The rounding off of the corner at the high-temperature end and the similar 'tail' in the specific-heat curve (fig. 20) are probably due to slight impurities in the specimen rather than to the irregular shape, for as can be seen from fig. 13 for instance, an irregular shape can usually cause only a gradual beginning of the transition, but not a gradual end.

Keesom and van Laer (1938), and in it are shown several lines of constant H_e for a spherical specimen. As long as the temperature is so low that $H_e < \frac{2}{3}H_c$, the sphere will be wholly superconducting with entropy S_s given by the lower curve; for $H_c < H_e < \frac{2}{3}H_c$, the sphere will be in the intermediate state with an entropy

$$xS_n + (1-x)S_s$$

where x, the fraction of the specimen in the normal state, is given by $3(H_e/H_c) - 2$, and finally, when $H_e > H_c$, the specimen will be in the normal state, with entropy S_n given by the upper curve.

With the help of fig. 22 it is easy to deduce the magneto-caloric cooling; thus if the sphere is thermally isolated at temperature A (2·6° K.) and a magnetic field greater than $\frac{2}{3}H_c$, say 165 gauss, is applied, the sphere will have to cool down to B (2·2° K.), in order that its entropy should remain constant. Evidently, for a given initial temperature, the final temperature will be lowest if the field applied is greater than the critical field at the final temperature reached, e.g. starting from A, the lowest temperature would be C (2·0° K.), which could be obtained by applying a field greater than 210 gauss. We may notice that if the applied field is not greater than the critical field for the final temperature, the final temperature will be less for a specimen of greater demagnetizing coefficient, since the entropy curve in the intermediate state is then more spread out (to put it another way, more of the specimen will be in the normal state for a given external field than for a specimen of smaller demagnetizing coefficient).

It will be seen from fig. 22 that for tin, starting with the lowest ordinarily available temperature, say 1° K., the lowest temperature that can be obtained by applying a magnetic field is something like 0·1° K. Still lower temperatures could, of course, be obtained by means of a cyclic process, in which the superconductor cooled down another superconductor, not in a magnetic field, which could then be separated from the first and magnetized, starting already from the lower temperature, and so on. The idea of using this cooling effect as a method of obtaining very low temperatures was first proposed by Mendelssohn and Moore (1934), but although they, and also Keesom and Kok (1934c), have shown that a lowering of temperature can indeed be obtained, the method has not yet been developed for practical use.* We may mention that it has several disadvantages compared with the Debye-Giauque method of adiabatic demagnetization of paramagnetic salts.

The most serious of these is the very low absolute value of the specific heat of the metal, so that, if it is used to cool down some other substance of comparable or greater specific heat, the drop of temperature produced will be much reduced. This can be seen at once from fig. 22, for if some other substance of greater specific heat is added to the metal, the entropy curves of the whole system will become much steeper, so that there will be much less difference between the temperatures of the system for which the entropy is the same with the metal superconducting and normal. Thus the method would be useless for cooling down any appreciable amount of liquid helium or paramagnetic salt, and would be inefficient even for cooling down some other non-superconducting metal. Its main use, in fact, would be to cool down the superconductor itself, for the purpose of studying its properties (for instance, the entropy-temperature relation itself) at low temperatures, and then it has the obvious advantage of requiring only relatively small magnetic fields. We should mention one other feature of the method, which tends to reduce its efficiency; this is that in order to keep the process adiabatic, the magnetic field must be applied infinitely slowly. Indeed, during the cooling process, the specimen is in the intermediate state, so that the change of the field induces eddy currents in the normal regions, whose dying out is accompanied by the irreversible production of Joule heat; consequently, unless the increase of field is made very slowly,† the temperature drop produced will be less than that calculated on the assumption of constant entropy.

* Daunt and Mendelssohn (1937) have pointed out that niobium and tantalum might be very suitable substances for the purpose of reaching low temperatures, since the entropy differences for these metals are relatively high (on account of their relatively high critical fields); see also Mendelssohn, 1952.

† An order of magnitude calculation based on kinetic arguments by Pippard (1950d) shows that for a 1 cm. diameter specimen the change of field would have to take some minutes in order that the gain in entropy from this cause should be negligible.

3.4. Calorimetric behaviour of alloys

If H_c and dH_c/dT for alloys were really as high as suggested by resistance measurements (§2.8) they should show an enormous specific heat jump. As we should expect from the evidence already presented in §2.8, no jump has been found in the specific-heat temperature curve of anything like the order of magnitude originally expected from the critical fields suggested by the resistance measurements (Mendelssohn and Moore, 1935b; Schubnikow and Chotkewitsch, 1934). This is, of course, because the bulk of the

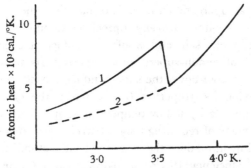

Fig. 23. Temperature variation of the specific heat of the alloy Sn+4% Bi: (1) without, and (2) with, a frozen-in field (Mendelssohn, 1935).

metal has, as we have seen in §2.8, quite normal critical fields, so that the specific heat of the whole specimen should behave in much the same way as for a pure element. The experiments were designed to look only for the originally expected large jump in the specific heat, and were not sensitive enough to show any small changes in the specific heat; probably a more accurate experiment would show actually not a sharp discontinuity in the specific-heat curve, but merely a hump over the temperature range in which the transition between superconductivity and normal conductivity takes place in the case of an alloy.

The calorimetric method can be used to show that after a sufficiently large magnetic field has been applied and removed from a superconducting alloy (i.e. when the specimen is left with a frozen-in flux), the bulk of the specimen is indeed in the normal state. Thus Mendelssohn and Moore (see Mendelssohn, 1935) measured the temperature-dependence of the specific heat of an impure tin specimen (which as regards the absence of the Meissner effect behaves like an alloy), first when it had been cooled in zero magnetic field, and then when a large field had been applied and removed at the lowest temperature. Their results, shown in fig. 23, show that in the second case the specific heat follows the curve corresponding to the normal state without any appreciable discontinuity, i.e. that the frozen-in flux prevents nearly all of the specimen from becoming superconducting again. With greater sensitivity, this method (which is somewhat analogous to Keesom and Kok's method of determining the proportions of normal and superconducting regions during the transition of a pure specimen in a field) might perhaps be used to find how much of the volume of the alloy was occupied by the super-conducting meshes, when there was a frozen-in flux in zero field.

3.5. Thermodynamics of mechanical effects

So far we have been investigating the thermodynamics of the superconducting transition regarding the pressure p and the volume V of the metal as constants; actually, however, we saw in §1.4.3 that there is a slight dependence of the critical field on stresses as well as on temperature, and we shall now deduce the thermo-dynamical consequences of this dependence. We shall deal only with uniform pressures, since owing to the meagreness of the experimental data, there is little point in considering any more complicated stresses.

Since the volume is now to be regarded as a variable, it is convenient to specify it more definitely as $V_s(H_e)$ in the superconducting state or V_n in the normal state, and we then have for a long rod parallel to the applied field

$$G_s(H_e) = G_s + \frac{H_e^2}{8\pi} V_s(H_e), \qquad (3.13)^*$$

and the fundamental equation (3.1) becomes

$$G_n = G_s + \frac{H_c^2}{8\pi} V_s(H_c). \qquad (3.14)$$

---

* The second term on the right-hand side is not quite correct, since in the integration on which it is based, the variation of V_s with H_e (see (3.16)) has been ignored; it is, however, easy to show that the error involved is negligible.

Since $(\partial G/\partial p)_T = V$ we find that

$$V_n - V_s(0) = V_s \frac{H_c}{4\pi}\left(\frac{\partial H_c}{\partial p}\right)_T + \frac{H_c^2}{8\pi}\left(\frac{\partial V_s}{\partial p}\right), \qquad (3.15)$$

but we have also from (3.13) that

$$V_s(H_c) - V_s(0) = \frac{H_c^2}{8\pi}\left(\frac{\partial V_s}{\partial p}\right), \qquad (3.16)$$

so it is evident that the last term of (3.15) merely represents a magnetostriction of the superconductor, i.e. a change of size which occurs in the purely superconducting state due to the pressure of the applied field. We shall see below that this last term is in fact negligible compared with the first. We thus have, if we now use V_s to denote the volume in the superconducting state at the critical field,

$$V_n - V_s = V_s \frac{H_c}{4\pi}\left(\frac{\partial H_c}{\partial p}\right)_T. \qquad (3.17)$$

If we substitute $\quad \left(\frac{\partial H_c}{\partial p}\right)_T = -\left(\frac{\partial H_c}{\partial T}\right)_p\left(\frac{\partial T}{\partial p}\right)_{H_c}, \qquad (3.18)$

and remember that $-V_s \frac{TH_c}{4\pi}\left(\frac{\partial H_c}{\partial T}\right)_p$ is the latent heat Q, we notice that (3.17) reduces to just the ordinary Clausius-Clapeyron formula

$$(\partial p/\partial T)_{H_c} = Q/T(V_n - V_s), \qquad (3.19)$$

where $(\partial p/\partial T)_{H_c}$ means the rate of change of pressure with temperature required to keep the constant applied magnetic field just critical.

If now we differentiate (3.17) with respect to T and p respectively, we obtain expressions for the differences of thermal expansion and compressibility in the normal and superconducting phases. Thus, ignoring second-order terms, we have

$$\frac{1}{V}\left(\frac{\partial V_n}{\partial T} - \frac{\partial V_s}{\partial T}\right) = \frac{1}{4\pi}\frac{\partial H_c}{\partial T}\frac{\partial H_c}{\partial p} + \frac{H_c}{4\pi}\frac{\partial^2 H_c}{\partial p\,\partial T}, \qquad (3.20)$$

and $\quad \frac{1}{V}\left(\frac{\partial V_n}{\partial p} - \frac{\partial V_s}{\partial p}\right) = \frac{1}{4\pi}\left(\frac{\partial H_c}{\partial p}\right)^2 + \frac{H_c}{4\pi}\frac{\partial^2 H}{\partial p^2}. \qquad (3.21)$

Since few data exist about the second-order pressure derivatives of H_c,* and since, moreover, we shall be concerned only with orders

* The values of $\partial^2 H_c/\partial p\,\partial T$ for tin and indium may be estimated from the data of Table II; the variation of $\partial H_c/\partial p$ with p is less than the error of the experiment, i.e. within experimental accuracy H_c varies linearly with p.

of magnitude, we shall consider only the special forms assumed by
(3.20) and (3.21) for $H_c=0$, i.e. in the absence of a magnetic field.
Rewriting (3.20) and (3.21) in terms of the coefficient of cubical
thermal expansion α, and the bulk modulus κ, where

$$\alpha = (\text{I}/V)(\partial V/\partial T) \quad \text{and} \quad \kappa = -V\partial p/\partial V,$$

we then find for $H_c=0$

$$\Delta\alpha = \frac{\text{I}}{4\pi}\frac{\partial H_c}{\partial T}\frac{\partial H_c}{\partial p}, \tag{3.22}$$

$$\Delta\kappa = \frac{\kappa^2}{4\pi}\left(\frac{\partial H_c}{\partial p}\right)^2. \tag{3.23}$$

If we use (3.18) these reduce to Ehrenfest's well-known relations
for a transition of the second kind (Ehrenfest, 1933)

$$\frac{dT_c}{dp} = VT_c\frac{(\alpha_n-\alpha_s)}{(C_n-C_s)} = \frac{\kappa_n-\kappa_s}{\kappa^2(\alpha_n-\alpha_s)}.$$

Until recently the only experimental data on the effects of
pressure on H_c and T_c were those of Sizoo and Onnes (1925) and
Sizoo, de Haas and Onnes (1925), which gave little more than
an indication of the order of magnitude of $\partial H_c/\partial p$.* Much more
detailed data have recently been obtained by Lasarew and Kan
(1944a, b) and Kan, Sudovstov and Lasarew (1948, 1949), who
applied high pressures by means of an ingenious ice-bomb technique in which water contained in a bomb is frozen round the
specimen; the pressure could be determined from the change of
size of the bomb. They found that for tin, for a pressure of
1.75×10^9 dyne cm.$^{-2}$ (about 1750 atmospheres), the transition
temperature was reduced by $0.1°$ K. and that H_c was reduced by
about 14 gauss (consistently with (3.18) if $(\partial T/\partial p)_{H_c}$ is assumed to
have its value for $H_c=0$). Thus $\partial H_c/\partial p \sim -10^{-8}$ gauss dyne^{-1} cm.^{2}
and $\partial T_c/\partial p \sim -5 \times 10^{-11}$° K. dyne^{-1}cm.^{2}; similar results were
obtained for other superconducting metals (see Table II), but it
should be noted that the sign of the effect is not always the same.†

* In the first edition of this book a numerical slip was made in the figure
quoted for this quantity, so that all the numerical conclusions deduced there are
invalid. This was pointed out by Lasarew and Kan (1944b).
† Experiments have also been made on the influence of tension on the
properties of tin (Sizoo and Onnes, 1925; Alekseyevsky, 1945a) and tantalum
(Alekseyevsky, 1940). For these metals H_c and T_c are increased by tension.

TABLE II

(a) Values of $10^{11} \times \partial T_c/\partial p$ (° K. dyne^{-1} cm.^{2})

Tl[3]	In[2]	Sn[2]	Hg[3]	Ta[3]	Pb[2]	NiBi$_3$[1]	RhBi$_4$[1]	Au$_2$Bi[1]
1·2	−4·7	−5·8	<0	<0	< −3	0·6	∼0·3	<0

(b) Values of $10^9 \times \partial H_c/\partial p$ (gauss dyne^{-1} cm.^{2})[2]

$T°$ K.	3·7	3·3	3·0	2·5	2·15
In	—	7·5	7·1	6·4	6·0
Sn	8·8	7·7	7·4	6·3	—

Notes. For Hg, Ta, Pb, RhBi$_4$ and Au$_2$Bi only qualitative statements are made in the published papers; for Tl the figure quoted is little more than an order of magnitude, since the result was very sensitive to the method of mounting the specimen.

References. (1) Alekseyevsky (1949a). (2) Kan, Sudovstov and Lasarew (1948). (3) Kan, Sudovstov and Lasarew (1949).

Substituting into (3.17) we find that for low temperatures $(V_n - V_s)/V_s \sim -10^{-7}$ (we may note here that the second term of (3.15) due to magnetostriction gives a relative volume change of only 3×10^{-9} at $2°$ K., i.e. about 30 times less). Quite recently Lasarew and Sudovstov (1949) have succeeded in measuring this small change of volume directly, using a bimetallic strip of tin and a non-superconducting metal (brass). As can be seen from fig. 24, their results are in good agreement with the prediction of (3.17) using the detailed data on H_c and $(\partial H_c/\partial p)$ at various temperatures.

The order of magnitude of $\partial T_c/\partial p$ is of some theoretical interest, for if it is interpreted as a change of T_c with the change of the crystal lattice cell dimension, say l, due to the pressure, we find for tin

$$\frac{l}{T_c}\frac{\partial T_c}{\partial l} \sim 30,$$

which is surprisingly large, for it means that if we put

$$T_c = T_{c0}(l/l_0)^n,$$

n has to be of order 30. The significance of this in Heisenberg's theory of superconductivity will be mentioned in §6.4.

If we substitute $\partial H_c/\partial p \sim 10^{-8}$, $\partial H_c/\partial T \sim 140$, $\kappa \sim 10^{12}$ in (3.22) and (3.23) we find

$$\alpha_n - \alpha_s \sim 10^{-7}, \quad (\kappa_n - \kappa_s)/\kappa \sim 10^{-5}.$$

The change of compressibility is probably too small to detect, and this is consistent with the negative results of de Haas and Kinoshita (1927), but the change of thermal expansion coefficient is quite comparable with the value of α_n and should be detectable with sufficiently delicate methods. The only attempt at detection has been with lead by McLennan, Allen and Wilhelm (1931 a) who could, however, only just detect α_n itself close to the transition temperature, so that it is difficult to decide whether a discontinuity as large as α_n itself was present or not.* They also looked for changes

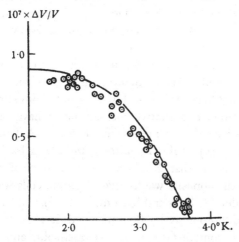

Fig. 24. Temperature variation of $\Delta V/V$ for tin (Lasarew and Sudovstov, 1949) The points shown are from series of measurements, in between some of which the apparatus was dismantled and reassembled. The full curve is calculated from (3.17) using the data of Kan, Sudovstov and Lasarew (1948).

of size when superconductivity was destroyed in a magnetic field, but could say only that $\Delta V/V$ was less than 3×10^{-8} at $4 \cdot 2^\circ$ K.; this result appears to contradict the observation of Kan, Sudovstov and Lasarew (1948) that $\partial H_c/\partial p$ is higher for lead than for tin (no figure is quoted) so that a value of $\Delta V/V$ of at least 4×10^{-7} should have been found. The negative result was probably due to the fact that a transverse magnetic field was used, so that the change of volume would have been gradual rather than sudden, and so might have escaped notice.

* See also Westerfield (1939) for a discussion of these data.

3.6. Thermal conductivity

The variation of thermal conductivity of metals with temperature is complicated by the fact that there are several different mechanisms of conduction, which are of varying relative importance according to the particular metal, its purity and the temperature. Moreover, these different mechanisms are differently affected by the transition to the superconducting state, so it is not possible to speak of any unique 'ideal' behaviour which is characteristic of all super-conductors, as we could for the magnetic properties for instance. In this section we shall describe the facts and classify them as far as possible in terms of the different mechanisms known to be relevant for metals in the normal state.

The older experiments of de Haas and Bremmer (1931, 1932a, 1936) and Bremmer and de Haas (1936a) gave a valuable indication of some of the qualitative features, but it is only comparatively recently that accurate quantitative measurements have become available for a variety of pure metals (de Haas and Rademakers, 1940; Rademakers, 1949; Hulm, 1949, 1950; and Mendelssohn and Olsen, 1950a). The fact that the thermal conductivity of certain alloys behaves in the opposite way to that of pure metals was first discovered in Leiden (de Haas and Bremmer, 1932b), and has been investigated more thoroughly by Bremmer and de Haas (1936b), Mendelssohn and Pontius (1937) and Mendelssohn and Olsen (1950a, b, c). The complicated nature of the phenomena is best illustrated by figs. 25–28, which show the temperature variation of the thermal conductivity κ, for pure tin, mercury and lead and for a typical alloy (lead + 10 % bismuth). The curves marked N are those for the normal state, produced by applying a sufficiently large magnetic field.* One important feature shown by all these curves is that there is no discontinuity in κ at the normal transition temperature; the theoretical significance of this will be mentioned in § 6.3.6 (see footnote, p. 215). In the early experiments any small discontinuity would have been smeared out by the temperature differences of order a few tenths of a degree used in the measure-

---

* Hulm (1950) found that for pure tin there was a considerable increase of thermal resistance with magnetic field in the normal state; the data plotted have been extrapolated to zero field.

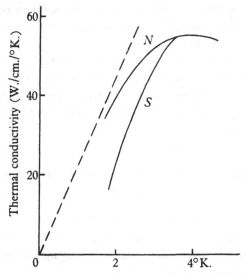

Fig. 25. Temperature variation of thermal conductivity of tin (Hulm, 1950). The broken line represents $L_0 T/\rho$ (see (3.25) and (3.26)).

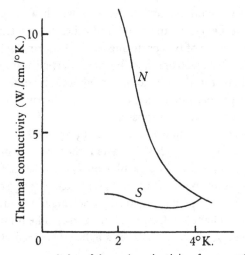

Fig. 26. Temperature variation of thermal conductivity of mercury (Hulm, 1950).

ment, and it is only recently that the absence of discontinuity has been established with any precision, by limiting the temperature differences to one- or two-hundredths of a degree (by application of a sensitive differential gas-thermometer technique (Hulm, 1950).

Before discussing the interpretation of the curves, it is necessary first to give a brief summary of the theoretical predictions about thermal conductivity in normal metals at low temperatures. In his survey of the problem Makinson (1938) found that (subject to certain limitations) the conductivity can be expressed as the sum of two terms: one, κ_e due to electronic conductivity, and the other, κ_g due to conductivity by the lattice, i.e. a conductivity which would occur in an insulator. Thus

$$\kappa = \kappa_e + \kappa_g. \tag{3.24}$$

The electronic thermal resistance $1/\kappa_e$ is made up of two parts:

$$1/\kappa_e = 1/\kappa_{eg} + 1/\kappa_{ei} = \alpha T^2 + \beta/T, \tag{3.25}$$

corresponding to the processes of scattering by the lattice vibrations (αT^2) and by impurities and imperfections (β/T) respectively; β is related directly to the residual electrical resistivity ρ_0, by the well-known Lorenz relation

$$\beta = \rho_0/L_0 \quad \text{with} \quad L_0 = \pi^2 k^2/3e^2, \tag{3.26}$$

while α is given by a complicated expression, which we shall not need, involving the Debye temperature and other parameters. As for κ_g, this too is made up of two parts due to the scattering of lattice waves by the metallic electrons and by other lattice waves; the detailed expressions for κ_g are complicated and involve many simplifying assumptions, but fortunately this qualitative description will be sufficient for our purposes.

For pure metals the κ_e term is dominant in (3.24), and Hulm has shown that the equation (3.25) fits the facts rather well for pure or only slightly impure metals. Thus a plot of T/κ against T^3 gives a reasonably straight line whose intercept gives excellent agreement with (3.26), and whose slope α is practically unchanged for addition of impurities (which change β drastically of course). The numerical value of α is, however, of order 10 times smaller than predicted by Makinson.

For tin and indium αT^2 is small compared with β/T; for mercury, on the other hand, it is the dominant term except at the lowest temperatures, while for lead the two terms are comparable. This is due simply to the fact that the Debye temperature of mercury is a good deal lower than that of tin and of indium, while that of lead

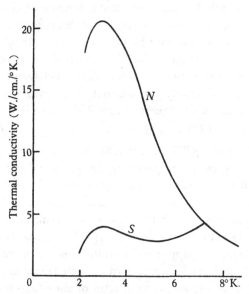

Fig. 27. Temperature variation of thermal conductivity of lead
(de Haas and Rademakers, 1940).

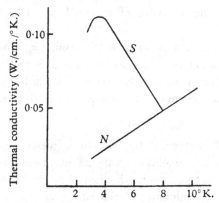

Fig. 28. Temperature variation of thermal conductivity of
Pb + 10 % Bi alloy (Mendelssohn and Olsen, 1950a).

lies in between, so that the effects of lattice vibrations are relatively
more important for mercury and lead. This difference between the
basic mechanisms for the thermal conductivity in tin and mercury
led Hulm to the suggestion that figs. 25 and 26 are really two
extreme forms of behaviour, characteristic respectively of scattering

of electrons dominantly by impurities and scattering of electrons dominantly by lattice vibrations. The results for lead (fig. 27) at low temperatures (and also some results of Rademakers (1949) on a specimen of tin appreciably purer than that used by Hulm) can then be thought of as representing an intermediate behaviour. The differences are more strikingly brought out by plotting the ratio κ_s/κ_n as a function of reduced temperature T/T_c. Thus if we put

$$\kappa_{eis}/\kappa_{ein}=f(T/T_c) \quad \text{and} \quad \kappa_{egs}/\kappa_{egn}=g(T/T_c), \qquad (3.27)$$

the forms of $f(T/T_c)$ and $g(T/T_c)$ are roughly as shown in fig. 29 by the tin and mercury curves respectively; lead nearly follows the mercury curve at high temperatures, where the lattice vibration mechanism is more important, and appears to continue the tin curve at low temperatures, where the impurity scattering is the more important. The suggestion that $f(T/T_c)$ and $g(T/T_c)$ are in some sense characteristic functions is supported by Hulm's results on less pure tin specimens, where the ratio κ_s/κ_n is hardly affected by an impurity content which increases κ_s and κ_n by a factor of more than 10. For still greater impurity contents the ratio curve begins to depart from the characteristic 'f' function, and this departure can very plausibly be associated with the influence of κ_g which we have hitherto ignored.

When in fact κ becomes as low as 0·1 W. units, κ_g is no longer negligible, and by a detailed analysis Hulm is able to find some evidence that κ_g (which is mostly due to scattering of lattice waves by electrons) is also influenced by the superconducting transition in such a way that

$$\kappa_{gs}/\kappa_{gn}=h(T/T_c), \qquad (3.28)$$

with $h(T/T_c)$ given by curve h in fig. 29. This suggests in turn that the anomalous behaviour of alloys (e.g. fig. 28) may be at least partly associated with the fact that the mechanism of thermal conduction in most alloys is predominantly of the lattice variety, since the residual electrical resistance is very high and hence κ_e much smaller than in a pure metal.

An interesting question about the functions f, g, h is whether they can be regarded as 'characteristic' in more than a qualitative sense; that is to say, whether they are independent of the metallic species. Hulm's measurements on indium and tantalum suggest that the function f though similar in type is probably not identical as between different metals; a definite answer is complicated by the uncertainties associated with the separation of κ_g from κ_e. Similarly, it does not look as if the function g is exactly the same for lead and mercury, though here again complete separation of g and f is difficult. An

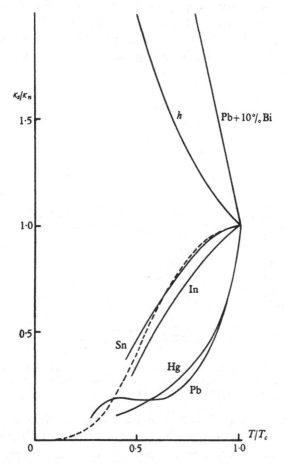

Fig. 29. Variation of κ_s/κ_n with T/T_c (Hulm, 1950; de Haas and Rademakers, 1940; Mendelssohn and Olsen, 1950a). The broken line is a plot of Heisenberg's formula (see equation (6.104)); the curve h is explained in the text.

attempt to change the mercury ratio curve from the g-type to the f-type by addition of impurity was not very successful owing to the complication of the κ_g term, which became important just in the relevant range of impurity concentration. Probably Hulm's interpretation should not be taken too literally, since it is in any case likely that the various mechanisms interact in a complicated way (which is ignored in the theory), but qualitatively it does provide a useful guide to what at first sight appears a hopelessly entangled situation. We shall return in §6.3.6 to the question of the possible theoretical interpretation of the functions f, g and h.

An indication that Hulm's interpretation of the anomalous behaviour of alloys is probably inadequate as it stands, is the fact

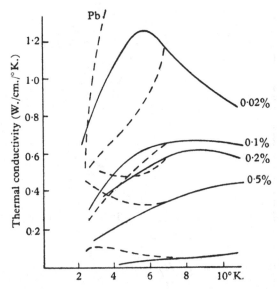

Fig. 30. Temperature variation of thermal conductivity of lead alloyed with various percentages of bismuth (Mendelssohn and Olsen, 1950*b*). Full curves: normal state; broken curves: superconducting state.

that the κ_s/κ_n-T/T_c curve found by Mendelssohn and Olsen for the lead-bismuth alloy is very much steeper than the $h(T/T_c)$ curve deduced by Hulm. A further indication comes from more recent experiments of Mendelssohn and Olsen (1950*b*), in which it is possible to follow how the inequality $\kappa_s < \kappa_n$ changes to $\kappa_s > \kappa_n$ as the bismuth concentration is increased (see fig. 30). The feature that is difficult to reconcile with Hulm's interpretation is that at sufficiently low temperatures the conductivities of some of the alloys in the superconducting state rise to values nearly as high as that of pure lead; the evidence on this point is not, however, quite conclusive, since the measurements do not reach to quite low enough temperatures and, moreover, the thermal conductivities of lead given by Mendelssohn and Olsen are a good deal lower than those

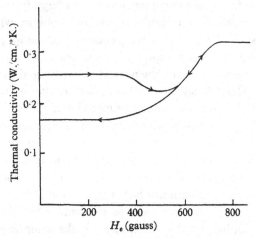

Fig. 31 a. Variation of thermal conductivity of the alloy Pb + o·1 % Bi with a transverse magnetic field at 2·7° K. (Mendelssohn and Olsen, 1950 a).

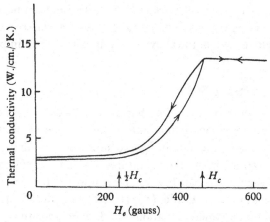

Fig. 31 b. Variation of thermal conductivity of lead with a transverse magnetic field at 4·85° K. (de Haas and Rademakers, 1940).

found by de Haas and Rademakers, suggesting that the lead used was not of such good quality.

An interesting feature of Mendelssohn and Olsen's (1950 a, c) results on the lead-bismuth alloys is the behaviour in a transverse

magnetic field (fig. 31 *a*). It can be seen that for an alloy with 0·1 %
bismuth the thermal conductivity is lower in the intermediate state
region than it is in either the normal or the superconducting state,
and it is lowest on the return curve when there is a large frozen-in
moment. These results (and similar ones for niobium which
behaves rather like an alloy) are in striking contrast to those for pure
lead (fig. 31 *b*), where the thermal conductivity varies monotonically
between the superconducting and the normal values.* On the basis
of their results, Mendelssohn and Olsen offer an alternative to
Hulm's interpretation of the thermal conductivity of alloys. They
suggest that the high thermal conductivity in the superconducting
state of alloys is due to a counter-flow mechanism analogous to that
which causes the high thermal conductivity of liquid helium, but
they have not yet worked out this idea in sufficient detail to show if
it can explain the facts quantitatively.† On this basis the lower
thermal conductivity in the intermediate state and when flux is
frozen-in is attributed to the hindrance of the counter-flow
mechanism by the fine structure of the intermediate state. It might,
however, be possible to interpret the lower thermal conductivity
as due to additional heat resistance at boundaries between normal
and superconducting regions, if it were supposed that the structure
of the intermediate state in the alloy is on a finer scale than in pure
lead.

3.7. Thermoelectric effects

Many experiments have shown that all thermoelectric effects
disappear in the superconducting state. The absence of electromotive force $\mathscr{E}$ in a circuit containing two superconducting metals
whose junctions are at different temperatures was first demonstrated
by Meissner (1927), and had been since confirmed by Borelius,
Keesom, Johansson and Linde (1931), Keesom and Matthijs
(1938*a*) using potentiometer methods, and more sensitively by

---

* If thermal resistance rather than thermal conductivity is plotted, the
variation in the intermediate state region is practically linear, which is to be
expected if the metal is broken up into normal and superconducting laminae
which lie across the axis of the cylindrical specimen (see Chapter IV). It is also
relevant to mention that the minimum in the thermal conductivity in the intermediate state of the alloy disappears at higher temperatures (e.g. 4·8° K. for the
specimen of fig. 31 *a*).

† This idea has also recently been discussed by Ginsburg (1950).

Burton, Tarr and Wilhelm (1935) and by Casimir and Rademakers (1947) using a superconducting galvanometer technique. The most sensitive demonstration of the vanishing was that of Steiner and Grassmann (1935) who suspended a superconducting circuit (both lead-tin and tin-indium were used) by a torsion fibre in a weak magnetic field; in the course of several hours they could detect no twist of the circuit when the junctions were unequally heated, and so were able to deduce that the thermoelectric power of the circuit must have been less than about 10^{-15} V./° K. (estimating an upper limit for the possible value of $L\,di/dt$ that might have escaped notice). This experiment provides a nice example of the improbability of existence of a thermoelectric force in the superconducting state. If a thermal e.m.f. had really existed the current would have grown without limit (or more precisely until superconductivity was destroyed), which *a priori* would seem to be unlikely.

Using the same experimental arrangement, Steiner and Grassmann induced a persistent current in the lead-tin circuit, and deduced from the fact that the current stayed constant throughout the experiment that the Peltier effect must also have been zero. This, of course, merely confirms the well-known Thomson relation for the Peltier coefficient Π_{AB}

$$\Pi_{AB} = T d\mathscr{E}_{AB}/dT, \qquad (3.29)$$

where $\mathscr{E}_{AB}$ is the thermoelectric force in a circuit of A and B whose hot junction is at temperature T. The second Thomson relation*

$$\sigma_A - \sigma_B = T d^2\mathscr{E}_{AB}/dT^2, \qquad (3.30)$$

shows that the Thomson coefficients σ of all superconducting metals must be equal, and since Nernst's theorem predicts that they must vanish at 0° K., it is very plausible to expect them to vanish over the whole superconducting range. This has been confirmed experimentally by Daunt and Mendelssohn (1938, 1946), who showed by means of a persistent current in an unequally heated lead circuit that $\sigma < 4 \times 10^{-10}$ V./°K.

Although there is no dispute about the disappearance of all thermoelectric effects in the superconducting state, there is a good deal of confusion about the behaviour of the thermoelectric e.m.f.

* It may be noticed that the objection to the usual elementary proof of the Thomson relations, that irreversible heat is associated with the transport of charge in a metal, does not arise in a superconducting circuit.

both in the region just above T_c, and below T_c in magnetic fields big enough to destroy superconductivity. In order to appreciate the position it is necessary to consider in some detail the experimental arrangements used and the results obtained.

Since there is no e.m.f. in a circuit of two different super-conductors, it follows that in a circuit of a normal metal and a super-conductor the e.m.f. is the same whatever superconductor is used. Thus the 'absolute thermoelectric power' e of a metal can be defined* as the value of $d\mathcal{E}/dT$, where $\mathcal{E}$ is the e.m.f. of a circuit of the metal and a superconductor with junctions at temperatures T and T_0; e for any superconductor is of course zero. The e.m.f. for a circuit made up of a number of metals $A, B, ..., N$ with junctions at temperatures $T_{AB}, T_{BC}, ..., T_{NA}$ is given by the integral with respect to T of e right round the circuit. If we denote $\int^T e_X \, dT$ by $\epsilon_X(T)$ for any metal X, the total e.m.f. can be written as

$$\mathcal{E}_{AB...N} = \epsilon_A(T_{NA}) - \epsilon_A(T_{AB}) + \epsilon_B(T_{AB}) - \epsilon_B(T_{BC})$$
$$+ ... + \epsilon_N(T_{MN}) - \epsilon_N(T_{NA}). \quad (3.31)$$

(note that it is essential that each metal should have completely homogeneous properties wherever the temperature changes). It is, of course, possible to add a different arbitrary constant to the ϵ's of each metal, without influencing the value of $\mathcal{E}_{AB...N}$, since each such constant appears twice with opposite signs. For definiteness it is convenient to define the ϵ's so that $\epsilon(0) = 0$. In principle it is possible that the thermoelectric powers may also depend on an applied magnetic field H. No simple expression can then be given for $\mathcal{E}$ unless H varies only where the temperature is uniform; if this is the case the value of the uniform field over each region of varying temperatures must be specified, i.e. if $\epsilon_X(T)$ is specified more precisely as $\epsilon_X(H, T)$, then (3.31) still applies.

We shall now consider two main types of experimental arrangement which have been used. The first of these is shown schematically in fig. 32: A is the metal to be studied, the temperatures T and T_0 being chosen so that in the magnetic field H, the upper part of A is normal and the lower superconducting, with a 'junction' at the temperature T_H for which H is the critical field of A. The metal

---

* See Borelius, Keesom, Johansson and Linde (1932).

B is either a non-superconductor (e.g. copper), or a superconductor with such high critical fields that it remains superconducting in all fields used (e.g. lead). The e.m.f. developed across the open ends in the situation shown is

$$\mathscr{E}(H) = \epsilon_{AN}(H, T) - \epsilon_{AN}(H, T_H)$$
$$- \epsilon_B(H, T) + \epsilon_B(H, T_0); \quad (3.32)$$

no contribution has been written for the superconducting portion, since ϵ has the same value for all points of a super-conducting metal. If H is increased beyond the critical field of A at tem-perature T_0, the situation is modified since the whole of A is then normal, and instead of (3.32) we have

$$\mathscr{E}'(H) = \epsilon_{AN}(H, T) - \epsilon_{AN}(H, T_0)$$
$$- \epsilon_B(H, T) + \epsilon_B(H, T_0). \quad (3.33)$$

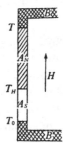

Fig. 32. Schematic diagram illustrating method used by Keesom and Matthijs (1938a), Casimir and Rademakers (1947) and Steele (1951). The suffixes N and S indicate normal and superconducting states.

This relation can be used to test whether or not the thermoelectric power of a metal depends on H. In fact, Steele (1951) showed with an arrangement of this kind (with B copper and A lead, tin, indium or thallium) that $\mathscr{E}'(H)$ did not vary appreciably with H. Since this result was found for various metals A it implies that all the ϵ's may after all be regarded as independent of H (except of course that $\epsilon(T_H)$ depends on the temperature at which H is the critical field) and the discussion is much simplified. Although it is very plausible that H should have no effect, Keesom and Matthijs in earlier experiments, both by this (1938a) and another method (1938b), came to the opposite conclusion; before discussing this contra-dictory evidence, however, it is convenient to complete the discus-sion of Steele's experiments, on the assumption that H has no influence except in determining the temperature at which A changes from the normal to the superconducting state. It then follows from (3.32) that

$$\mathscr{E}(H) - \mathscr{E}(0) = \epsilon_{AN}(T_c) - \epsilon_{AN}(T_H). \quad (3.34)$$

Thus by measurements of *changes* of $\mathscr{E}$ with H (which also eliminate the inevitable stray e.m.f.'s in the measuring circuit), Steele was able to determine $\epsilon_{AN}(T_H)$. His result can be expressed as

$$\epsilon_{AN}(T) = \tfrac{1}{2}\alpha T^2, \quad (3.35)$$

(the arbitrary constant having been chosen to make ϵ_{AN} vanish at $T=0$), though as is evident from (3.34) his method cannot give any information about ϵ_{AN} above T_c. It follows from (3.30) that the Thomson coefficient of the normal metal A is

$$\sigma_{AN} = \alpha T. \tag{3.36}$$

On certain simplifying assumptions (see, for instance, Seitz (1940), and also Pimentel and Sheline (1949)) this is just the result predicted by the theory of metals applied to a free electron model; the theory of metals predicts also that the coefficient α should be of the order of magnitude $\frac{2}{3}\gamma/e$, where γT is the electronic specific heat per atom and e the electronic charge.* As can be seen from Table III, Steele's values of α are in good agreement with this prediction.

TABLE III

(See Steele, 1951)

Metal	$10^8 \times \alpha$, V./$(°$ K.$)^2$	$10^8 \times \frac{2}{3}\gamma/e$, V./$(°$ K$)^2$
Tl	1·9	0·8
In	1·8	1·0
Sn	0·8	1·1
Pb	2·2	2·0

This detailed discussion of Steele's results, which are of interest mainly for the theory of the normal state of metals, may seem out of place in a book primarily about superconductivity; the discussion is, however, relevant because of the contradictory results of other investigations which suggest that the thermoelectric behaviour of a metal close to the threshold curve already 'announces' the proximity of superconductivity. Keesom and Matthijs (1938 a) also used the arrangement of fig. 32, but with the further simplifications that B was superconducting lead, so that (3.32) simplifies to

$$\mathscr{E}(H,T) = \epsilon_{AN}(H,T) - \epsilon_{AN}(H,T_H), \tag{3.37}$$

and in particular

$$\mathscr{E}(0,T) = \epsilon_{AN}(0,T) - \epsilon_{AN}(0,T_c), \tag{3.38}$$

while (3.33) becomes

$$\mathscr{E}'(H,T) = \epsilon_{AN}(H,T) - \epsilon_{AN}(H,T_0). \tag{3.39}$$

The measurements were carried out somewhat differently from those of Steele, inasmuch as H was kept constant and T varied. To

* It is assumed that there is one free electron per atom.

eliminate stray e.m.f.'s in the measuring circuit, the e.m.f. of the circuit in any conditions was always measured as the difference between $\mathscr{E}(H, T)$ (or $\mathscr{E}'(H, T)$) and $\mathscr{E}'(H, T_0)$ (this last should of course vanish, if it were not for the stray e.m.f.'s in the circuit, which were actually quite large). If ϵ_{AN} is independent of H (except through T_H), as found by Steele, the curve of $\mathscr{E}(H, T)$ against T for fixed H and for $T > T_c$ should be displaced by the constant amount $\epsilon_{AN}(0, T_c) - \epsilon_{AN}(H, T_H)$ from the curve $H = 0$. Keesom and Matthijs found this not to be the case for indium. Moreover, the curve of $\epsilon(0, T)$ for indium (which of course can be followed only down to T_c) has a rather rapid change of slope in a range of about $1°$ K. above T_c, as if some feature of superconductivity had begun to appear well above T_c.

Casimir and Rademakers (1947) have made measurements on a tin-lead circuit, with the same basic scheme, but with a much more sensitive method of measuring the e.m.f.'s (essentially a superconducting galvanometer). They found that for tin also, $\epsilon(0, T)$ has a sharp change of slope, but this time only about $0.15°$ K. above T_c (see fig. 33); they made no measurements in a magnetic field which can be compared with those of Steele.

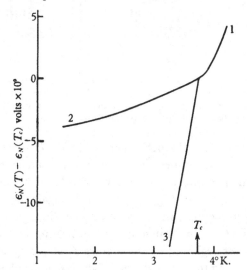

Fig. 33. Temperature variation of $\epsilon_N(T) - \epsilon_N(T_c)$ for tin. (1) Casimir and Rademakers (1947). (2) Keesom and Matthijs (1938b). (3) Webber and Steele (1950) and Steele (1951).

Keesom and Matthijs (1938 b) have also used a somewhat different technique of measurement, and a very similar method has more recently been used by Webber and Steele (1950); the method is illustrated in fig. 34. The junctions to metal B are now at the same temperature, but part of the metal A is screened from the magnetic field H by the lead cylinder L; in fig. 34 a H is large enough to maintain the normal state at $T = T_0$. There must therefore be two junctions of normal to superconducting metal A—one well inside the zero field region at $T = T_c$ and the other at the mouth of the lead cylinder in a field which has the critical value for temperature T_0. It is easily seen that the e.m.f. is now

$$\mathscr{E}(H) = \epsilon(H_c(T_0), T_0) - \epsilon(0, T_c). \tag{3.40}$$

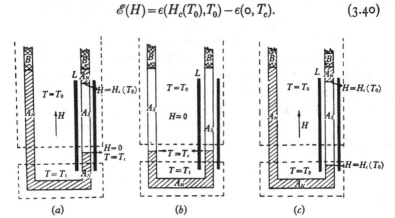

Fig. 34. Schematic diagram of method used by Keesom and Matthijs (1938 b) and Webber and Steele (1950). In each case the broken lines indicate the boundaries of isothermal enclosures, and L indicates the lead cylinder; the suffixes N and S indicate normal and superconducting states.

For $H = 0$ (fig. 34 b) a second junction appears at $T = T_c$ in the other arm of the loop, while the upper junction disappears and so there is no e.m.f. If in the presence of H the upper and lower regions are both at the same temperature (fig. 34 c) the e.m.f. is again zero. Thus the stray e.m.f.'s in the measuring circuit may be eliminated either by switching off H, or by switching off the heating coil which maintains the lower region at a temperature T_1 above T_0, and in either case subtracting the residual e.m.f. (which may be identified as the 'stray' e.m.f.). Keesom and Matthijs adopted the latter technique, while Webber and Steele in their repetition of this type of experiment adopted the former procedure. The results of Keesom and Matthijs for indium by this method agreed fairly well with their earlier measurements by the other method,[*] but agree only very roughly with the data of Steele already described. For tin the agreement with Steele is even worse; this is illustrated in fig. 33, which shows also Casimir and Rademakers's results for $T > T_c$. Webber and Steele's repetition of the lead-cylinder technique, however, gives very good agreement with the results of Steele for

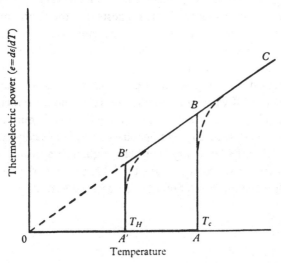

Fig. 35. Schematic diagram illustrating 'ideal' temperature variation of thermo-electric power. $OA'B'C'$ indicates behaviour in a magnetic field H, and $OABC$ in the absence of a magnetic field; the broken curves indicate schematically the observed departures from ideal behaviour. The line OC is drawn straight on the basis of Steele's data (1951).

tin. In view of the reasonable order of magnitude of Steele's results and the good agreement with Webber and Steele, they are probably more reliable than the others.

If we ignore the quantitative discrepancies between different authors, the situation may most simply be discussed in terms of a temperature plot of the thermoelectric power $e = d\epsilon/dT$ as shown in fig. 35. The most plausible behaviour is that indicated by the

---

[*] They confirmed also the field dependence of ϵ, by measurements on an all-normal circuit, part of which was screened from the field H. Their description, however, does not make perfectly clear what exactly was measured.

straight lines, $OB'BC$ representing the normal state, and $OA'A$ the superconducting state, with a sharp drop to zero from one to the other at the appropriate temperature. The results of Keesom and Matthijs and Casimir and Rademakers are illustrated schematically by the curves departing from the 'ideal' scheme. It is natural to suppose that just as for the resistance-temperature relation, these departures are due to non-ideal conditions. Indeed, it is improbable that if a magnetic field really affects $e(T)$, and if there is really an anomalous region just above T_c, these effects should vary so much from one experiment to another. This suggests that the effects may be due either to some feature of the measuring technique (for instance, the Wood's metal ring seal in the apparatus of Keesom and Matthijs might on account of its superconductivity distort the applied magnetic field) or to impurities which produce inhomogeneities in regions of varying temperature. It should be remembered also that all the arguments have assumed a sharp boundary between normal and superconducting phases, while actually it is more probable that the boundary is either irregular or spread over a large area in the presence of a magnetic field; this would almost certainly modify the interpretation of the observed e.m.f.'s.

CHAPTER IV

STRUCTURE OF THE INTERMEDIATE STATE

4.1. Introduction

We have seen in Chapter II that for any configuration other than a long cylinder in a uniform magnetic field parallel to the axis, destruction of superconductivity by a magnetic field takes place over a *range* of magnetic fields, and, moreover, in this intermediate-state region the specimen must break up into some kind of mixture of the normal and superconducting states. In this chapter we shall discuss the structure of this mixture for specimens which can be treated as ellipsoids (or approximately so), leaving out of consideration the more complicated types of intermediate state which occur in non-ellipsoidal specimens (see § 2.7). We shall also consider the problem of destruction of superconductivity by a current in a wire (§ 4.7) which involves a somewhat different kind of intermediate state.

This branch of the subject has had a curious historical development, inasmuch as the most direct experimental evidence for the structure of the intermediate state came only *after* Landau had developed theories on the basis of less direct (and sometimes unreliable) experiments. In our presentation we shall follow a logical rather than a historical order, though referring to the historical aspects where they are of interest. To make clear the relevance of the various experiments we shall first give an outline of the theoretical position (§ 4.2), but only the general ideas and the final formulae will be given, since the details of the mathematical development are rather complicated and can be found in the original papers. This will be followed by an account of the direct evidence from the beautiful experiments of Shalnikov and Meshkovsky that the metal in the intermediate state is indeed broken up into a complicated mixture of pure superconducting and pure normal regions (§ 4.3), and then the older evidence coming from resistance measurements will be discussed (§ 4.4).

The theory involves a characteristic length Δ (connected with the surface energy at the phase boundary between normal and superconducting regions), and indicates that even when the specimen size is as much as 100 times Δ, the properties of the specimen should be size-dependent. This idea suggested that some of the mysterious departures from macroscopic predictions in the resistance measurements were due to the small diameter of the wires used. This was confirmed in a series of experiments by the author and colleagues on wires and plates in transverse magnetic fields in which the size was deliberately varied and the resistance and magnetic moment studied (§§ 4.4, 4.5). Other evidence for the existence of a surface energy between the normal and superconducting phases comes from the phenomenon of 'supercooling', a feature of the magnetization curve which has been ignored in Chapter II. Recent experiments by Faber which have produced a better understanding of this effect will be described in § 4.6.

4.2. Theoretical considerations

We have already seen in § 2.5 that the destruction of superconductivity in an ellipsoid by a magnetic field cannot take place by a simple spreading of a normal region from the outside of the specimen, since this leads to the paradoxical situation of normal regions existing at places where the field is less than H_c. Now we have seen in Chapter III that the critical field has a simple thermodynamic interpretation as the field in which the free energies of the superconducting and normal phases become equal, so that instead of saying that it is 'paradoxical' to have normal regions where the field is less than critical, we could equally say that the free energy is not as low as it would be if part of the normal regions were superconducting. In fact, the real problem is to find an arrangement of superconducting and normal regions such that the free energy of the whole specimen is as low as possible.

Well inside the specimen, the boundaries must presumably run parallel to the uniform applied field H_c, as illustrated in fig. 36, since if any boundary were curved the field would either be less than H_c in a normal region, or exceed H_c at a superconducting surface. At the entries and exits of the flux, however, difficulties would arise with the model of fig. 36, owing to the excess free energy associated with the sharp corners, where the field must certainly exceed H_c. This excess free energy can be reduced by suitably rounding the corners, but a compromise is inevitable, since if they are rounded as illustrated in fig. 36, the boundary being chosen so that the field is just H_c on it, then clearly the field is again *less* than H_c in the normal regions close to the curved portions, while if there is a field higher than H_c at the boundary, the superconducting regions have not their lowest free energy.

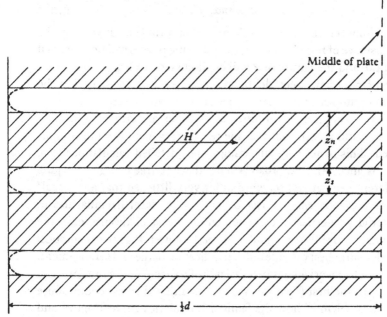

Fig. 36. Schematic diagram illustrating a possible structure of a plate in the intermediate state. The curved lines represent schematically boundaries at which the field is just critical.

Leaving aside for the moment the question of how best the compromise should be made, it can be seen that the total free energy associated with the best compromise can be lowered by increasing the number of regions into which the specimen is divided. Roughly speaking this is because the lines of force coming from outside are the less distorted by the presence of superconducting regions, the closer together are the normal regions between them. Evidently it would be profitable to make the subdivision infinitely fine if it were not for the existence of a positive surface free energy at the inter-phase boundary; the finer the division the larger is the area of contact between the two phases, and so the larger is the unfavourable balance of surface free energy. The free energy difference between superconducting and normal phases per unit volume is $H_c^2/8\pi$, as has been shown in §3.1, so if the surface free energy per unit area is α, it will be seen that the theory must involve a characteristic length Δ given by

$$\Delta = 8\pi\alpha/H_c^2. \qquad (4.1)^*$$

The importance of α and Δ for other problems involving the coexistence of normal and superconducting phases will be discussed in §4.6, and the nature of α will be discussed in §6.3.5.

So far we have been deliberately vague about the shapes of the regions into which the specimen is divided in the intermediate state. The two simplest possibilities to consider are either laminae or threads, but it turns out that the real arrangement is more complicated. This is not very surprising, since detailed calculations (see, for instance, Andrew, 1948 b) show that the free energies of quite different geometrical schemes differ very little, so that any 'ideal' scheme can be easily upset by very slight local irregularities of material properties. For concreteness we shall discuss only the laminar model, but it should not be forgotten that the theoretical results can at best provide only an indication of the real arrangement. Landau had worked out two variants of a theory of the intermediate state. In the first (1937) it was assumed that the best compromise in the problem of how the laminar boundaries curve at entry and exit is the one shown in fig. 36, with the boundaries determined by the condition that the field should be exactly H_c on them. This assumption is, however, to some extent arbitrary, since it is not proved, and it is indeed probably not true, that this arrangement guarantees the smallest free energy. In the second variant (1938, 1943) Landau pointed out that in his 1937 theory he had not completely avoided the difficulty of having normal regions where the field was less than H_c (close to the convex entry and exit regions), and proposed an ingenious new solution to avoid the difficulty. This is to branch the normal laminae as shown in fig. 37, in such

---

* The notation used here is that of Pippard (1951 a); it should be noted that the effect of slight field penetration into the superconducting phase is included in α. Thus our α is the α'_{ns} of Ginsburg (1945) and our Δ the β'_{ns} of Ginsburg, and the Δ' of Désirant and Shoenberg (1948 a).

a way that there are no places where the boundaries have appreciable curvature, until close to the surface the division is so fine that the laminae are thinner than the penetration depth and the flux emerges quite uniformly over the whole surface. Thus close to the surface the division is supposed to be so fine that no distinction can be made between the two phases and a new kind of state, the 'mixed state', occurs.

We shall see in §4.3 that the existence of this uniform 'mixed state' at the surface has been disproved by Shalnikov and Mesh-kovsky's experiments which have shown that distinct superconducting and normal regions come right out to the surface. Indeed, it has never been proved that the branching model has a lower free energy than the other type of model, and the experimental facts suggest that Landau's earlier variant is nearer the truth. The mathematics of the earlier variant is more difficult than that of the branching model, and so the properties of the non-branching model have not been as fully worked out as those of the branching one, but recently Kuper (1951) has made calculations on a non-branching model using a method which involves less specific assumptions and which is more readily applied to practical problems.

In the working out of both the Landau models the special case of a thick plate of infinite area transverse to the field is first considered. For this case $n = 1$, and we have simply $B = H_e$ and $H = H_c$ in the intermediate state (see (2.17)). Thus, since practically all the flux must be carried by the normal laminae, it follows that in either model the ratio of the thicknesses of the normal to superconducting laminae at the middle of the plate is very close to

$$z_n/z_s = B/(H_c - B) = H_e/(H_c - H_e). \qquad (4.2)$$

The main problem is now to work out the geometry of the structure and to find how z_s, and the period z (given by $z = z_n + z_s$) of the structure, depend on d, the plate thickness.

In Landau's non-branching model, the shape of the laminar boundaries close to the specimen surface is worked out on the assumption that the field must be exactly H_c on it (which, as we have already mentioned, is not necessarily the correct assumption), and then the resulting free energy of the specimen for the geometry is minimized with respect to z. For the branching model it is again assumed that the field is exactly H_c in the laminae (though for specimens of finite thickness this assumption proves to be only approximately true). The detailed geometry of the branching shown in fig. 37 is determined first by choosing the angle θ so that for a given splitting of a normal lamina the sum of the extra surface free energy (inversely proportional to θ) and the extra volume free energy (directly proportional to θ) shall be a minimum. The value of x is then chosen so that the laminae branching at the appropriate

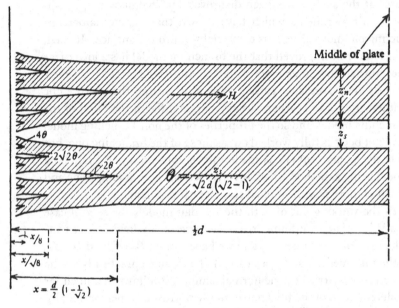

Fig. 37. Landau's branching model of the intermediate state
(Landau, 1938, 1943).

angles eventually just fill out the whole free surface of the specimen. The total free energy of this arrangement is then minimized with respect to z. Kuper's calculation of the non-branching model starts with a long cylinder of diameter d in a transverse field. The superconducting domains in the intermediate state are assumed to be circular disks, and their distorting effects on the field are calculated by considering their demagnetizing coefficients, treating them as flat ellipsoids of the same volume. In this way a value of z_s is obtained which minimizes the free energy; although the calculation really applies only to a transverse cylinder, it is plausible to suppose that the results still apply as regards order of magnitude to a transverse plate, and to other specimen shapes.

Denoting H_e/H_c by η, the final answers for the various estimates of z_s/d are as follows:

(1) Landau's non-branching model:

(a) H_e small ($\eta \ll 1$):

$$\frac{z_s}{d} = \left(\frac{\pi}{\ln(1/2\eta)}\right)^{\frac{1}{2}} \frac{1-\eta}{\eta} \left(\frac{\Delta}{d}\right)^{\frac{1}{2}}. \tag{4.3}$$

(b) H_e near critical ($\eta \to 1$):

$$\frac{z_s}{d} = \left(\frac{\pi}{\ln 2}\right)^{\frac{1}{2}} \left(\frac{\Delta}{d}\right)^{\frac{1}{2}}. \tag{4.4}$$

(No explicit formula can be given for the general case.)

(2) Kuper's non-branching model:

$$\frac{z_s}{d} = \frac{1}{(2\eta)^{\frac{1}{2}}} \left(\frac{\Delta}{d}\right)^{\frac{1}{2}}. \tag{4.5}$$

(3) Landau's branching model:

$$\frac{z_s}{d} = (2\sqrt{2}-2)^{\frac{2}{3}} \left(\frac{1}{\eta}-1\right)^{\frac{1}{3}} \left(\frac{\Delta}{d}\right)^{\frac{1}{3}}. \tag{4.6}$$

In spite of the apparent differences between the predictions of the various calculations, in practice they do not give very different results, although (4.5) gives values systematically lower than (4.3) and (4.4). To give an idea of the orders of magnitude, consider the numerical example of $d = 4$ cm. and $H_e/H_c = 0.8$ and assume $\Delta = 4 \times 10^{-5}$ cm. (which we shall see in §4.5 is the right order of magnitude), then (4.4), (4.5) and (4.6) give respectively $z_s \sim 0.28$, 0.10 and 0.5 mm. We shall see in §4.3 that 0.2 mm. is indeed the order of magnitude of the linear dimensions of the superconducting domains in the intermediate state as actually found by experiment.

The next stage in the theory is to apply it to determine the magnetic properties of the transverse cylinder; for the branching model this has been done by Landau (1943) and Andrew (1948b). Essentially the idea is to suppose that the structure is the same as that deduced for a plate, and then to determine H by differentiating the appropriate free energy with respect to B. Owing to the presence of the surface-energy terms, H is no longer exactly the same as H_c, so that I is no longer exactly given by (2.17)* and the differences will evidently depend on the specimen size. Kuper's calculations on the non-branching model start out with a transverse cylinder, so the magnetization curve is more easily obtained; his results are qualitatively similar to those obtained with the branching model, though there are quantitative differences.

One important feature of the departure from the macroscopic magnetization curve may be understood qualitatively in the following way. For a sufficiently large macroscopic specimen, the intermediate state sets in as soon as the field reaches H_c anywhere on the surface, since then the free energy of the intermediate state becomes equal to that of the superconducting state. This is only true, however, if the extra free energy due to the internal surfaces separating normal and superconducting regions is neglected, and, indeed, the larger the specimen the more is this neglect justified. Actually, however, if Δ/d is not completely negligible, the internal surface energy must be taken into account, and its effect is to postpone entry into the intermediate state to a slightly higher field than is predicted by macroscopic theory. Only then is there a sufficient balance of free energy available to create the new internal surfaces corresponding to the appropriate structure of the intermediate state. We have seen (§ 2.5) that the macroscopic theory predicts entry into the intermediate state at a field H_e given by

$$\rho = H_e/H_c = 1 - n,$$

(e.g. $=\frac{1}{2}$ for a transverse cylinder). Landau's branching model predicts that more accurately we should have for a transverse cylinder of radius r

$$\rho = \tfrac{1}{2} + 1\cdot28(\Delta/r)^{\frac{2}{3}}, \tag{4.7}$$

assuming a laminar structure, and

$$\rho = \tfrac{1}{2} + 2\cdot31(\Delta/r)^{\frac{1}{2}}, \tag{4.8}$$

assuming a thread-like structure (note that in Landau's (1943) paper there is a numerical error in the values of coefficients; the correct

---

* It should be noticed that there are two unsatisfactory features about this type of treatment: (1) the cylinder is treated really as a plate of variable thickness, and an average value of d has to be put into the equations for a plate; the choice of this value is somewhat arbitrary; and (2) although the field in the laminae comes out as different from H_c by an amount depending on specimen size, this is not taken into account in deciding the form of the branching or the laminar dimensions. It is not at all obvious that this approximation of putting $H = H_c$ at the beginning of the calculation is justified.

coefficients are taken from Andrew (1948 b)). Kuper's (1951) calculation gives

$$\rho = \tfrac{1}{2} + 0\cdot 42(\Delta/r)^{\frac{1}{4}}. \qquad (4.9)$$

Assuming $\Delta = 4 \times 10^{-5}$ cm. we see from any of these formulae that even for r as large as $0\cdot 4$ mm., differences in ρ of order 10 % are to be expected from the macroscopic theory values. We shall see in §§ 4.4 and 4.5 that this provides the solution to what was for many years known as the '$0\cdot 58$ mystery'—the fact that resistance in a transverse wire was restored not exactly at $\tfrac{1}{2}H_c$ but at $0\cdot 58H_c$. The more recent experiments designed to test the theory have shown also that the experimental values of ρ are indeed size-dependent.

Landau did not go beyond the calculation of ρ, but Andrew and Kuper have calculated the whole of the magnetization curve for different values of Δ/r; it will, however, be convenient to postpone further discussion of these theoretical predictions until the experimental results have been described in § 4.5.

4.3. Direct experimental evidence for the structure of the intermediate state

As already mentioned, the most direct evidence for a discontinuous structure in the intermediate state comes from a series of experiments by Shalnikov (1945), Meshkovsky and Shalnikov (1947 a, b) and Meshkovsky (1949). These experiments were very much influenced by Landau's branching model theory, and particularly by his prediction that no discontinuous structure could be expected at the surface of an ellipsoid owing to the intervention of the mixed state. Thus, it seemed futile to look for a structure at the outside surface of an ellipsoid, and instead the aim was to look for the structure in a narrow gap between two halves of the specimen. Indeed, Landau had calculated that if a sphere were cut into halves separated by a narrow gap, the mixed state would not intervene unless the gap exceeded a critical size (of order 10^{-2} cm. for a sphere of 2 cm. radius), so that the kind of field distribution at the middle of the plate of fig. 37 should be observable in a gap narrower than the critical size.

After some only partially successful attempts to reveal the field distribution in the gap by means of a ferromagnetic powder, Shalnikov (1945) made an ingenious application of the bismuth-wire technique. At the time he did not consider it possible to use a bismuth wire short enough to be able to resolve the expected fluctuations of the field over the surface of the gap, so he used instead a long bismuth wire and made use of the non-linear properties of the bismuth wire as a field detector. The resistance R of the wire is approximately proportional to the square of the field, and the resistance of a length of wire should therefore be proportional to the average of the square of the field over the length of the wire rather than to the square of the average field. Now the average field is B, while according to the theory the actual field is zero immediately over superconducting regions and H_c over normal regions. Thus, if the calibration of the wire in a uniform field is used, the wire will appear to be in a field greater than B when the sphere is in the intermediate state. In fact the fraction of the wire covered by normal patches is $x = B/H_c$, so the root-mean-square field should be $(BH_c)^{\frac{1}{2}}$, which is greater than the mean field B. The experiment showed that if the gap was fairly wide the field measured was indeed B, giving the variation with H_c shown in fig. 11a, but when the gap was very narrow the intermediate state part of the curve was bowed, giving field values higher than indicated in fig. 11a, and thus proving that the field was really not uniform. The result for a wide gap was taken at the time as confirmation of the critical gap prediction, but we shall see below that this conclusion was later proved unsound.

These early experiments are now only of historical interest as having been the first to demonstrate a discontinuous structure, for they have been entirely superseded by the even more direct experiments of Meshkovsky and Shalnikov (1947a, b) and Meshkovsky (1949). In these the bismuth wire was made so small that it became possible to reveal the discontinuous changes of field as sudden changes of resistance for movements of the wire close to the superconductor. It soon became apparent that there was really no critical gap width at all; if the gap was wide and the probe was moved about in a plane not sufficiently close to the metal surfaces, the field appeared more uniform simply owing to a 'fringeing' effect. Thus, the field variation along a line parallel to the metal surface changes progressively with distance from the metal surface in the manner indicated schematically in fig. 38. The experiments

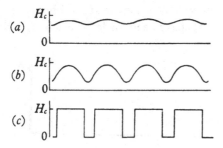

Fig. 38. Schematic illustration of the smoothing out of field variation parallel to the surface of a plate divided into normal and superconducting laminae, (a) along a line relatively far from the surface, (b) along a line closer to the surface, (c) along a line very close to the surface.

made it quite clear, however, that even for very wide gaps the discontinuous structure was still present close to the surface. In other words there is no evidence at all for the mixed state assumed at the surface in the branching model.

Once this was clear, the original objection to studying the field distribution at an outer surface no longer existed, and Meshkovsky and Shalnikov (1947b) have confirmed that the same discontinuous structure may be observed over the outside of the sphere. This is illustrated in fig. 41, which may be compared with fig. 39, showing the field variation in the gap between two hemispheres; from a detailed survey in such a gap, Meshkovsky (1949) has constructed the topographical 'maps' shown in figs. 40(a) and (b). The following points may be noted:

(1) The structure is very complicated—there is, for instance, evidence for small superconducting 'islands' inside normal material, so that the field issuing out of normal patches has not always the full H_c value. Clearly neither a laminar nor a thread model can be regarded as an adequate description.

(2) The size of the superconducting domains, for $H_e/H_c = 0.8$, is of order 0·2 mm. for a tin sphere of radius 2 cm. at 3° K., though it is difficult to determine any definite size precisely.

Such a size would be expected on the non-branching model for $\Delta \sim 10^{-4}$ cm. (this is based on equation (4.5), which presumably applies very roughly if the radius of the sphere is substituted for d).

(3) The structure is rather different according as the point (H_e, T) in the intermediate state is reached by increasing H_e at

constant T or reducing T at constant H_e. Although no strictly comparable experiments have been made in which H_e is decreased

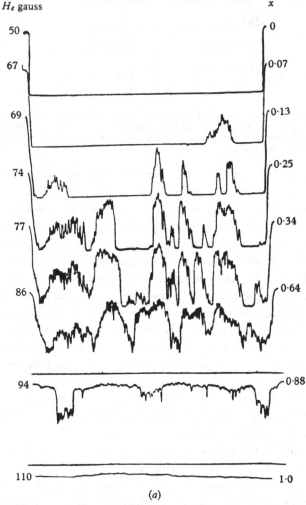

(a)

Fig. 39. Field variation across diameter in a gap 0·12 mm. wide between two tin hemispheres (4 cm. diam.) (Meshkovsky and Shalnikov, 1947a). (a) for increasing fields at $T = 3°$ K. (b) (see opposite page) for decreasing temperatures at $H_e = 72$ gauss.

at constant T, or T increased at constant H_e, it is probable that the relevant factor is not whether H_e or T is varied, but merely, whether the experimental conditions are approached from the direction of

destruction or creation of superconductivity. It is evident in fact
that the mechanism of entry into the intermediate state is different
according as the conditions are approached from the all-normal or

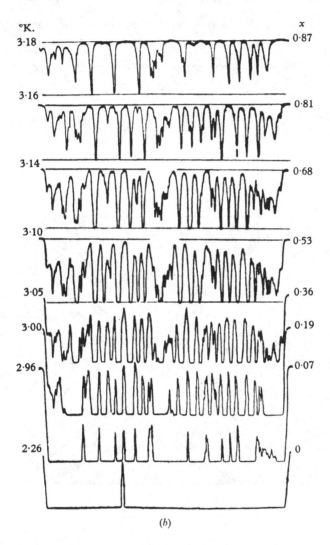

(b)

the all-superconducting state. This complication is not envisaged
by any of the existing theories which consider a unique equilibrium
state.

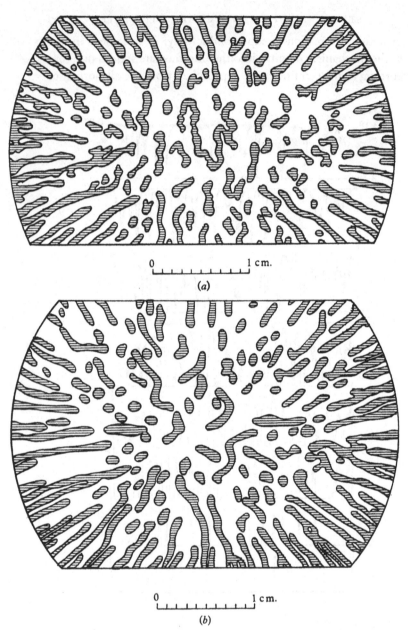

0 |_|_|_|_|_|_|_|_|_| 1 cm.

(a)

0 |_|_|_|_|_|_|_|_|_| 1 cm.

(b)

Fig. 40. Topography of field over gap between two tin hemispheres (Meshkovsky, 1949). Shaded areas represent normal regions, and white areas superconducting regions; the gap width was 0·2 mm. and the diameter was 4 cm. (a) For $H_e = 70$ gauss, $T = 3°$ K. (H_e increased at constant T to reach this point). (b) For $H_e = 81$ gauss, $T = 2·85°$ K. (T reduced at constant H_e to reach this point).

(4) The details of the structure were found to be extremely stable if H_e and T were held constant for some time. For instance, fig. 42 a

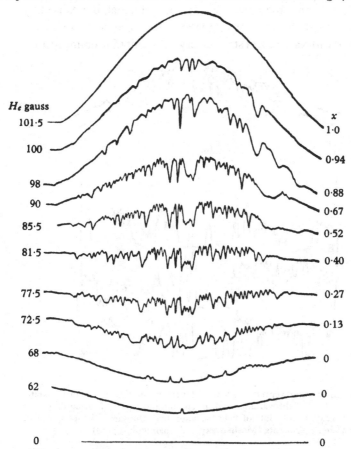

H_e gauss

101·5

100

98

90

85·5

81·5

77·5

72·5

68

62

0

x

1·0

0·94

0·88

0·67

0·52

0·40

0·27

0·13

0

0

0

Fig. 41. Field variation over the surface of a tin sphere for diminishing values of the applied field at $T = 2·97°$ K. (Meshkovsky and Shalnikov, 1947b). As the bismuth probe is moved over the surface of the sphere its orientation with respect to the field, and therefore its sensitivity, changes; the top curve illustrates the behaviour in a uniform field (sphere in the normal state). When the sphere is all superconducting the curve dips down because the field falls off away from the equator but the orientation effect is absent, since the field is then tangential to the surface. The interpretation at intermediate fields is therefore complicated, but qualitatively the presence of a structure is evident.

shows two graphs of field against position taken along the same line at times 20 min. apart, and it can be seen that all the details are still reproduced; while fig. 42b shows that even if the measurements were taken on quite different days (the specimen having been warmed up in between) many of the features are still reproduced. This suggests that these features are not accidental, but caused by the inevitable slight variations of material properties in the specimen associated with variation of stresses, crystal orientation or impurities.

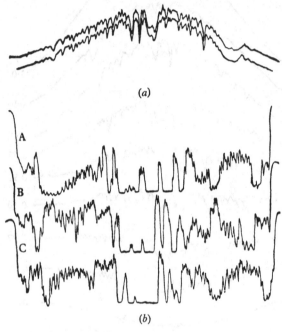

(a)

(b)

Fig. 42. Illustrating reproducibility of structure in the intermediate state. (a) Two plots of field variation taken 20 minutes apart (Meshkovsky and Shalnikov, 1947b). (b) Plots of field variation over the same diameter, but in three separate experiments (Meshkovsky and Shalnikov, 1947a).

It is to be hoped that these very interesting experiments will be continued in order to investigate other shapes and sizes of specimen and also to study the temperature variation of the scale of the structure (nearly all the experiments so far have been at a single temperature).

4.4. Resistance measurements

The idea that in the intermediate state the metal contains a mixture of normal and superconducting regions had already been suggested by resistance measurements on wires in transverse magnetic fields long before the direct evidence described in §4.3 was available. In fig. 43 we reproduce some experimental curves showing how the resistance of a cylindrical wire varies with the strength of the transverse field, and it will be seen that the resistance begins to return at approximately half the critical field, i.e. as soon as the cylinder enters the intermediate state. That this return of resistance at $\frac{1}{2}H_c$ is really connected with the commencement of penetration was shown by de Haas, Voogd and Jonker (1934) in an experiment with a cylinder of elliptical instead of circular cross-section, when, corresponding to the different demagnetizing

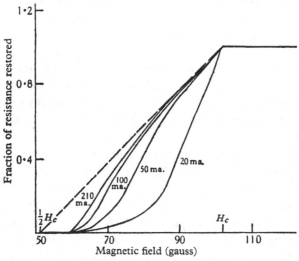

Fig. 43. Restoration of resistance of monocrystalline tin wire by a transverse magnetic field; diameter, 0·25 mm.; $T = 2·92$ °K. (de Haas, Voogd and Jonker, 1934).

coefficient, the resistance began to return at a different field value. It is interesting to mention that when von Laue (1932) first suggested that the return of resistance at $\frac{1}{2}H_c$ in the case of a circular cylinder in a transverse field was connected with the field being twice as great at the 'equator' of the cylinder cross-section as it would be in a longitudinal field, the idea of an intermediate state had not been introduced, and it remained a mystery why resistance should reappear if superconductivity was destroyed only at the very equator (for it seemed that a superconducting path would still be left open through the core of the cylinder).

This evidence that in the intermediate state there is a resistance which increases with increase of the external field (i.e. with increasing field penetration) does, indeed, fit in with the hypothesis that in the intermediate state the metal is divided up into microscopic superconducting and normal regions, arranged in such a way that a current has to pass partly through the normal regions. In order that there should be resistance to the current it would seem that the structure must be laminar rather than thread-like, and that the laminae must have their planes normal to the current direction. We may mention in this connexion an interesting experiment of Schubnikow and Nakhutin (1937) (for details see also Nakhutin, 1938), who showed that a sphere in the intermediate state has a resistance perpendicular, but not parallel, to the direction of the applied field, thus suggesting strongly that the sphere is indeed divided up into alternating superconducting and normal laminae. Less direct evidence of a similar kind was provided by the alternating field experiments with a sphere (Shoenberg, 1937b) which also suggested that the resistance of a sphere in the intermediate state was anisotropic.

This evidence for a laminar structure from the non-zero resistance perpendicular to the applied field is in apparent conflict with the evidence presented in §4.3, for a careful examination of the topography of normal and superconducting regions (e.g. fig. 40) suggests that a current could quite well find a completely superconducting path round the sphere without ever crossing a normal region. A clue to the solution of this contradiction is the strong dependence on measuring current of the resistance in the intermediate state. As can be seen from fig. 43, the resistance falls as the measuring current is reduced, and though there is no great consistency between the results of different investigations (de Haas, Voogd and Jonker, 1934; Misener, 1939; Andrew, 1948a), it is not impossible that in ideal conditions the resistance would become zero for vanishingly small currents. Similarly, in the alternating field experiments, the average resistance of the sphere was found to decrease markedly with the amplitude of the alternating field (i.e. with the strength of the eddy currents in the sphere), down to amplitudes as low as $3 \times 10^{-5} H_c$. This current dependence evidently indicates that the structure of the intermediate state is very sensitive to any current fed into the specimen from outside, and the resistance evidence might be reconciled with the somewhat sponge-like structure, found by Meshkovsky and Shalnikov, if it were supposed that this structure changes into a more laminar one as soon as a current is passed. An alternative interpretation, which does not, however, help resolution of the conflict with the Meshkovsky and Shalnikov evidence, is that for very low currents the laminae lie parallel to rather than across the current direction, though always parallel to the field direction.*

It should not be forgotten that so far the topography has been studied only for a sphere, and, moreover, only for a particular size, so it may yet prove that there is a different kind of topography for the transverse wire, on which most of the resistance evidence is based. Evidently it would be of interest to study directly how the field topography is influenced by passage of small currents, in, say, a wire transverse to a magnetic field. As we shall see in §4.5, there is some evidence that a measuring current has practically no influence on the magnetization of a transverse wire, but this merely suggests that the magnetization is not a very structure-sensitive property. The theoretical problem of current dependence has not yet been tackled; it was thought earlier that the mixed phase at the surface could be made responsible for some of the peculiarities of the current-dependent properties, but this hypothesis falls to the ground now that we know that there is no mixed phase.

Up to now we have left out of account what was for many years a most puzzling mystery, that the resistance of a transverse wire began not exactly at $\frac{1}{2}H_c$ but at $0.58H_c$ (this can be seen, for instance, in fig. 43). The solution of this '0.58 mystery' has already been indicated in §4.2 as due to a size effect, and the fact that it was always just about 0.58 that was found experimentally is explained by the fact that the wires used in the Leiden experiments were nearly always of about the same radius ($\sim 10^{-2}$ cm.). The first detailed investigation in which the size was deliberately varied was by Andrew (1948 a), and he did indeed find that ρ (defined as the

---

* It should be noticed that a lamina or filament lying parallel to the current direction is inherently unstable since the field will always be greater on one side of it than the other; this asymmetry would make a normal lamina grow in the direction in which the field exceeded H_c and shrink on the other side, i.e. the lamina would move sideways.

value of H_e/H_c for which resistance first appeared) could be made to vary from 0·67 to 0·55, as the radius increased from 1·4 × 10⁻³ to 53 × 10⁻³ cm.* and did not vary much with temperature.† By comparison of this variation with the theoretical formula (4.7) or (4.8), Δ for tin was estimated as 1·7 × 10⁻⁵ cm. Since, as we shall see in § 4.5, the magnetization curves of a transverse cylinder give practically the same value of ρ as the resistance measurements, but give, in addition, other interesting information, it will be convenient to postpone a fuller discussion of the ρ values to § 4.5. Certain other features of Andrew's results and of some resistance measurements on plates in a transverse field (Andrew and Lock, 1950) will also be discussed in § 4.5.

4.5. Magnetization measurements

As soon as Landau's theory of the intermediate state had shown that ρ should be size-dependent, it was evident that the 0·58 effect should not be a peculiarity of resistance measurements, but should show up also in magnetization measurements. In order to verify this and to test other aspects of the theory, magnetization curves of transverse cylinders of various radii were made by Désirant and Shoenberg (1948 a), using the method indicated in § 2.11.3.

A typical magnetization curve is shown in fig. 44, and fitted to it are theoretical curves as calculated by Andrew (1948 b) on the basis of the branching model and by Kuper (1951) on his non-branching model; it can be seen that the agreement between experiment and theory is tolerable. The following features may be noted:

(1) The wire remains wholly superconducting until a field appreciably higher than $\frac{1}{2}H_c$—for this specimen about 0·6H_c at 3·00° K. As in the resistance results, ρ increases as the specimen radius is reduced.

---

* The range of radii is limited at the upper end by the difficulty of measuring very low resistances accurately, and at the lower end by the difficulty of making fine enough wires.

† It should be mentioned that Andrew's results do not agree with earlier work of Misener (1938), who found that ρ varied strongly with temperature and approached $\frac{1}{2}$ at the normal transition temperature. It is, however, probable that Misener's results must be regarded as due to poor surface conditions, since Andrew found that he too could obtain results much like Misener's, in specimens whose surface had been contaminated.

(2) The magnetization drops rather sharply at entry into the intermediate state. In the theory this is associated with the fact that the free energy curves for the superconducting and intermediate states cross at a finite angle (instead of touching, as they do for a macroscopic specimen).

(3) The magnetization disappears at a field appreciably lower than H_c (as determined from measurements in a longitudinal field) and the difference becomes more marked as the radius is reduced.

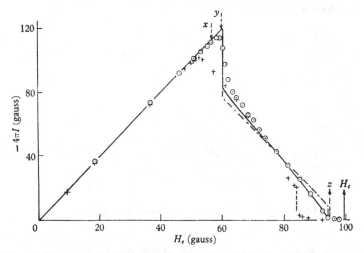

Fig. 44. Magnetization curve of cylindrical tin wire (radius $1 \cdot 5 \times 10^{-2}$ cm.) in transverse field at $3 \cdot 00°$ K. (Désirant and Shoenberg, $1948a$). ⊙ increasing fields; + decreasing fields. Full curve based on Landau's (1943) branching theory as worked out by Andrew ($1948b$) assuming $\Delta = 3 \times 10^{-5}$ cm.; chain curve based on Kuper's (1951) theory assuming $\Delta = 0 \cdot 5 \times 10^{-5}$ cm. x, first appearance of resistance; y, rapid rise in resistance; z, full resistance restored. Broken vertical line indicates discontinuous rise of $-4\pi I$ accompanied by discontinuous fall of resistance.

This is associated with the fact that the free energy per unit volume of the intermediate state is higher than it would be for a macroscopic specimen (owing to the internal surface energy), so the free energy of the normal state becomes equal to it at a somewhat lower field. Actually the theory suggests that the magnetization should disappear discontinuously (since the free energy curves of the intermediate and normal state cross), but this is not observed.

(4) On reducing the field from above the critical field, there is at first some slight hysteresis which can be associated with

supercooling (see §4.6), and then the main part of the curve in the intermediate-state region is retraced. This indicates that the magnetization is not very structure sensitive, since Meshkovsky and Shalnikov's results suggest rather different structures for the superconducting-normal and normal-superconducting transitions. The 'horn' part of the rising curve is not, however, retraced, suggesting either that this part is metastable or that the different structure on the return curve makes it impossible to expel all the flux at the foot of the 'horn'

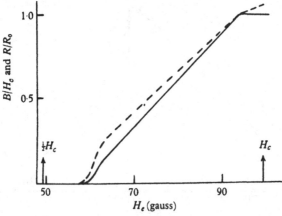

Fig. 45. Variation with applied transverse field of B/H_c (broken line) and R/R_0 (full line) at 3·00° K. for the specimen whose magnetization curve is shown in fig. 44; the current used for the resistance measurement was 30 mA. To avoid confusion only the curves for increasing fields are shown (Andrew, 1948 a).

In order to correlate the magnetization with the resistance data the resistance of the specimen whose magnetization curve is shown in fig. 44 was measured simultaneously with the magnetization in a special experiment. The result of this experiment is indicated in fig. 44 and illustrated more fully in fig. 45; it can be seen that the correlation is very close (compare features (1), (2), (3) above). Indeed, for a large measuring current the resistance curve is very nearly identical with a suitably scaled curve of B against H_e, showing again that in these conditions the normal regions (which occupy a fraction B/H_c of the whole volume) must lie continuously across the cross-section of the wire. For smaller measuring currents the resistance is smaller, presumably due to a change in this arrangement, but the magnetization is not noticeably affected. This points once again to the lack of structure sensitivity of the magnetization.

Detailed comparison of the magnetization with the branching model theory showed that although qualitatively the agreement was not too bad it was certainly not good in detail. Thus estimates of Δ from the ρ values, using (4.7), were several times lower than estimates based on fitting the descending part of the magnetization curve, while intermediate values of Δ were obtained from attempting to fit the whole theoretical curve as well as possible. The difficulty of deducing Δ from the experiments is that it enters into all the theoretical formulae in low powers such as cube or square roots, with the result that quite large changes of Δ can be made without altering the appearance of the theoretical curves very seriously. Similar difficulties, but in a more acute form, arise in the comparison of the data with Kuper's theory; his estimates of Δ based on the ρ values are systematically lower, while those based on the slope of the descending portion are systematically higher, than for the branching model. The magnetization data on any interpretation indicate a temperature dependence of Δ, as is shown by Table IV, where the various estimates of Δ are summarized.* We shall see in §4.6 that Faber's experiments on supercooling suggest a similar temperature variation of Δ. Owing to the inadequacies of the theories, not too much reliance can be placed on the absolute values of Δ, but it is probable that 10^{-5} cm. represents the right order of magnitude for tin at 2° K., and that Δ is rather smaller for mercury than it is for tin.

Since the basic theory of the branching model is based on the behaviour of a plate rather than a wire, Andrew and Lock (1950) made a series of measurements on the magnetization curves of thin disks in transverse fields. A typical curve is shown in fig. 46, and it will be seen that the 'horn' of the transverse wire (fig. 44) has now assumed grandiose proportions. Again, the branching theory fits

---

* The fact that Andrew's resistance measurements give rather less temperature-dependence of Δ is perhaps due to differences in the precise criteria for estimating ρ in the two sets of experiments; in the joint experiment described above, both resistance and magnetic measurements were in agreement in indicating a temperature variation of Δ. It should be noted that in any case the temperature variation of ρ found by Misener (1938) (see footnote, p. 114) is in the opposite sense to that found here.

qualitatively, but partly because it was possible to cover a much wider range of thicknesses than was possible for the wires, the quantitative discrepancies are now much more apparent. It is not possible to find a single value of Δ at a given temperature which will represent the height of the 'horn' at all satisfactorily for all thicknesses, and the whole form of the horn differs from the theoretical prediction. If, however, the horn is ignored, estimates of Δ based

TABLE IV. *Estimates of* Δ *(cm.* $\times 10^5$)

(a) *Transverse cylinders* (Désirant and Shoenberg, 1948a; Andrew, 1948a).

° K. ...	Sn			Hg			
	2·1	3·0	3·5	2·1	3·7	3·97	4·05
Landau branching theory:							
From ρ values (magnetization curves)	0·7	1·0	2·0	0·3	>0·3	>0·3	>0·3
From ρ values (resistance curves)	1·7	1·8	2·5	<0·5	<0·5	<0·5	<0·5
From slope of falling part of magnetization curve	2·0	3·6	6·2	1·2	2·1	2·8	2·1
From best overall fit of magnetization curve	0·8	1·5	3·7	0·4	0·7	0·9	1·1
Kuper theory:							
From ρ values (magnetization curves)	0·2	0·3	0·6	—	—	—	—
From slope of falling part of magnetization curve	~ 10	~ 20	~ 40	—	—	—	—

(b) *Transverse plates* (Andrew and Lock, 1950).

Landau branching theory:			
From slope of falling part of magnetization curve	—	3·7	6·3

Notes. Kuper's estimates of Δ from the falling part of the magnetization curve were made for each specimen separately, and scatter widely; the figures quoted are rough averages. In the analysis of the transverse plates, only one estimate is given, since it is clear that the theory does not fit well in other respects. Kuper's analysis of the transverse plate data gives very high values of Δ ($\sim 30 \times 10^{-5}$ cm.), based on the ρ values, but gives the form of the variation of ρ with thickness more correctly than the Landau theory.

on the descending part of the curve, agree surprisingly well with those obtained in the same way from the wires. This agreement suggests that perhaps the theory for this part of the curve is not very sensitive to the detailed assumptions of the branching model, i.e. that the true model will still give much the same size-dependence for this part of the curve. The application of Kuper's theory to the

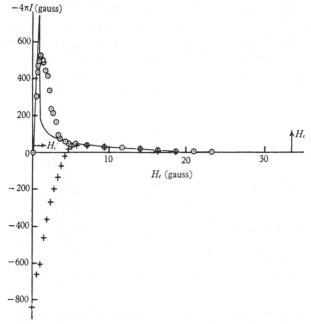

Fig. 46. Magnetization curve of tin disk (radius 0·37 cm., thickness 4·1 × 10⁻⁴ cm.) in transverse field at 3·49° K. (Andrew and Lock, 1950). ⊙ increasing fields; + decreasing fields. Full curve based on Landau's (1943) branching theory assuming $\Delta = 6\cdot3 \times 10^{-5}$ cm.

transverse plate can also account qualitatively for some features of the experimental data, but again leads to more serious quantitative discrepancies than the branching theory.

A clue as to why the theory should go wrong, particularly for the 'horn' part of the curve, lies in the irreversibility of this region. If formation of the intermediate state took place reversibly over the whole range of applied fields, the area under the I-H_e curve should always be $H_c^2/8\pi$, whatever the transition mechanism (see § 3.1), while the experimental curves have several times this area for the thinnest plates. Since it is just in the horn region that most of the extra area lies, it is evident that the theory (based on reversible processes) cannot be expected to fit this part of the curve. The strong frozen-in moments found in all these disks are probably associated with the edge effects discussed in § 2.7, and could in fact be appreciably reduced by electrolytic polishing, which smooths the edges. The nature of these frozen-in moments has been studied also by Alekseyevski (1946). Andrew and Lock (1950) also made experiments on the resistance of a thin plate (rectangular instead of circular, however), and the results are shown in fig. 47, together with the magnetization curve of a similar specimen. There is no longer the close correlation between resistance and magnetization results that was found for a wire, the resistance returning at a much

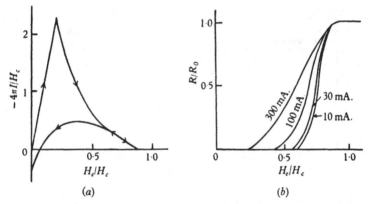

Fig. 47. (a) Magnetization curve, (b) resistance-field curve, for a rectangular tin plate (7·8 × 10⁻³ cm. thick and 0·1 × 0·73 cm. for (a) and 0·1 × 3 cm. for (b)) at 3·51° K. (Andrew and Lock, 1950).

higher field (strongly dependent on current) than that at which the magnetization starts to fall. This must mean that the normal regions in the early part of the intermediate state do not reach right across the path of the current flow, and shows that resistance in the intermediate state is by no means inevitable,

4.6. Supercooling

4.6.1. *Hysteresis due to supercooling* (see also § 5.2.1). We have up to now spoken as if there were no hysteresis in an 'ideal' magnetization curve (such as fig. 6a). Actually experiments on very pure specimens do show a hysteresis which becomes more marked the more 'ideal' is the specimen. This hysteresis (which is in-dicated by the broken lines in figs. 13 and 44) is in many ways reminiscent of the supercooling of a vapour below its liquefaction temperature, and we shall refer to it as 'supercooling', even though usually it is the magnetic field rather than the temperature which is varied. When the field is reduced from above H_c (or the tem-perature lowered in a steady field) the specimen stays in the normal state, until at a field appreciably less than H_c, a sudden expulsion of flux takes place and the specimen goes into either the intermediate state or, if it has a small enough demagnetization coefficient, directly into the pure superconducting state).

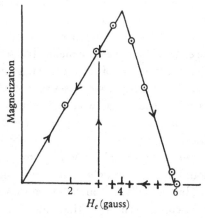

Fig. 48. Magnetization curve of an aluminium sphere at $1.16°$ K.* illustrating large supercooling effect (Shoenberg, 1940b). ⊙ increasing fields; + decreasing fields. The scale of magnetization was not calibrated.

In the early measurements with tin specimens only slight super-cooling was observed (of order 2 or 3 % of H_c) (Mendelssohn and Pontius, 1936a; Shoenberg, 1937a), but a much more marked effect has been found with pure aluminium (Shoenberg, 1940b), as illustrated in fig. 48, and quite recently Faber has been able to demonstrate marked effects in tin by using special techniques (see § 4.6.3). Similar effects had been observed also in the early Leiden measurements on resistance (Sizoo, de Haas and Onnes, 1925; de Haas, Sizoo and Onnes, 1925; de Haas and Voogd, 1928, 1931a),

* The temperature was originally given as $1.105°$ K., but a source of error has since been pointed out by Goodman and Mendoza (1951) and an adjustment to the 1949 scale has also been made.

but were influenced strongly by the method of attaching the potential leads and so were thought for some time not to be fundamental (see de Haas, 1933).

4.6.2. *Theoretical discussion.* H. London (1935) first suggested that such effects might be expected if there were a surface-free energy of suitable magnitude at the boundary between superconducting and normal phases, and that just as in the analogous problem of the vapour-liquid transition, supercooling occurs because of the difficulty of growth of a nucleus of the stable phase.

We shall first contrast briefly the consequences of a positive and a negative interphase surface energy. In fact, we have already seen that a positive surface energy has to be assumed to explain the features of the intermediate state in a pure metal, but as we shall see below there may be small local regions where an interface would have negative surface energy even in a pure metal. If the interphase surface energy is negative, the state of lowest free energy is one in which there is as much splitting into two phases as possible; thus in a field lower than critical the metal would consist of a superconducting matrix broken up by many very thin normal laminae, while above the critical field it would consist of a normal matrix broken up by many very thin superconducting laminae. Naïvely speaking 'many' should mean 'infinitely many' and 'thin' should mean 'infinitely thin', but actually, of course, the concept of surface energy loses its macroscopic meaning if the laminae become too crowded or too thin, and it is impossible to predict the detailed nature of the splitting without a detailed understanding of superconductivity. For our purposes the qualitative statement that splitting up would occur is sufficient; this splitting would presumably lead to abnormally low resistance persisting above the critical field, and it was just because this did not seem characteristic of 'ideal' conditions that H. London (1935) postulated the existence of a positive rather than a negative interphase surface energy.*

If the surface energy is positive, no superconducting phase should

---

* If the slight penetration of a magnetic field into the superconducting phase (see Chapter v) is alone considered it can be easily seen that this leads to a negative interphase surface energy $\lambda H_c^2/8\pi$, where λ is the penetration depth. As will be explained in § 6.3.5 this is however usually overcompensated by a positive energy associated with the gradualness of the configurational change between the two phases.

survive above the critical field, and no normal phase below (provided, of course, we are considering a body of zero demagnetizing coefficient so that the complications of the intermediate state do not intervene). The transition from normal conductivity to superconductivity must, however, start from some small superconducting nucleus in the normal matrix, and in order to understand why supercooling occurs we shall now consider the equilibrium of such a nucleus. Let its volume be V and its surface area A, and assume for simplicity that its demagnetizing coefficient is negligible; we shall suppose that the interphase surface free energy, α per unit area, may vary from place to place, but again for simplicity we shall assume that the values of α and its variation $\partial\alpha/\partial n$ along the normal to the surface of the nucleus are constant over the surface of the nucleus.* If the nucleus is to be in equilibrium at a particular size, the free energy of the specimen must remain unchanged for a small growth of the nucleus; if the free energy increases, the nucleus will shrink, while, if it decreases, the nucleus will grow. Thus, assuming that α is independent of the field, the conditions for growth, equilibrium or shrinkage of a nucleus are

$$\left(g_n - g_s - \frac{H^2}{8\pi}\right)\delta V \gtreqless \alpha\,\delta A + \frac{\partial\alpha}{\partial n}\,\delta V,$$

where g_n and g_s are the free energies per unit volume of the two phases (in zero field for the superconducting phase), and the upper inequality refers to growth. Substituting $H_c^2/8\pi$ for $(g_n - g_s)$ from (3.1) and using the notation $\Delta = 8\pi\alpha/H_c^2$ as already defined in (4.1), this becomes

$$\left(1 - \frac{H^2}{H_c^2}\right) \gtreqless \frac{\Delta}{d} + \frac{\partial\Delta}{\partial n}, \qquad (4.10)$$

where d is the appropriate small dimension of the superconducting nucleus defined by $d = \delta V/\delta A$; it should be noted that the precise meaning of d depends on the manner of growth considered—for a cylinder growing radially, d is just the radius, but for a cylinder

---

* It should be emphasized that these assumptions are essentially over-simplifications, since the shape of the nucleus should also be considered as a variable; thus for positive α the surface energy is least for a spherical shape, but the demagnetization energy is least for a long shape or a lamina. Moreover the concept of a well-defined surface boundary implied in the discussion is unlikely to be quite valid in a discussion of small regions in a non-uniform material.

growing longitudinally, d is half the radius. The nucleus will grow, shrink or be in equilibrium according as the left-hand side of (4.10) is greater than, less than, or equal to the right-hand side.

If Δ is everywhere positive it is clear that a superconducting nucleus must always collapse, however small H is, unless d can be brought up to a value such that the equality sign holds. Once the equality sign does hold, a further decrease of H (which increases the left-hand side) will tend to make the nucleus grow; this will reduce Δ/d but it does not follow that instability sets in, since if $\partial\Delta/\partial n$ increases sufficiently rapidly with outward movement of the boundary, it may be possible to maintain the equality sign in (4.10) with a larger value of d. Eventually, however, a stage will be reached where further growth makes the right-hand side smaller than the left, and the nucleus then becomes unstable; superconductivity spreads right through the metal and supercooling has broken down.

The fundamental problem that must now be considered is how a nucleus can be created of sufficient size to be able to grow further; unless such a nucleus is available, the superconducting phase can never grow and supercooling must continue to zero field. In the vapour-liquid transition an important mechanism for the creation of nuclei of sufficient size to grow is that of fluctuations, but in the present case there is only a negligible probability of a fluctuation creating a sufficiently large superconducting nucleus. This probability is proportional to $\exp(-\Delta G/kT)$, where ΔG is the Gibbs free energy excess of the nucleus; a detailed calculation of ΔG involves consideration of the shape of the nucleus, but it is easy to see that the order of magnitude of ΔG is not less than $(H_c^2 - H^2)\,V/8\pi$, where the volume V is that of a sphere of radius r given by

$$1 - (H/H_c)^2 \sim 2\Delta/r$$

(ignoring the fact that (4.10) does not apply strictly to a sphere, and ignoring the term in $\partial\Delta/\partial n$). Substituting 4×10^{-5} cm. for Δ and assuming a supercooling which gives $1 - (H/H_c)^2 = 0\cdot 1$, we find for $H_c = 100$ gauss, $T = 3°$ K. that $\Delta G/kT \sim 10^8$, so that the probability is indeed quite negligible.*

---

* More detailed considerations show that the constant of proportionality in front of the exponential, to give absolute probability per unit time per unit volume of metal, is not large enough to outweigh the exponential.

A more plausible mechanism for the appearance of a nucleus of sufficient size has recently been suggested by Faber (1952) on the basis of experiments which will be described below. The essence of his suggestion is that any real material can never be quite homogeneous, and if there are places where Δ is locally negative owing to inhomogeneous strains, they will act as superconducting nuclei. In fact, as already mentioned in §2.8, Lasarew and Galkin (1944) found that superconducting regions persist up to very high fields in specimens which have been inhomogeneously strained, and it is plausible to suppose that this happens because Δ is negative over finite regions of the material. A negative Δ over appreciable regions may also provide the explanation of the 'Mendelssohn sponge' (§2.8) in alloys; in alloys inhomogeneities are to be expected, and the 'meshes' of the sponge presumably occur in regions where Δ is negative. Above H_c, regions of negative Δ will split up in some way so as to retain superconducting inclusions even at very high fields (see p. 122); when the field is reduced below H_c these inclusions will expand until their boundaries come to places where the right-hand side of (4.10) is sufficiently positive to make (4.10) an equality. As already explained, such a nucleus will grow in a stable fashion as H is reduced until a stage is reached where further growth reduces the right-hand side of (4.10) and instability then sets in.

Faber has worked out this idea in greater quantitative detail, assuming Δ to vary in a particular way in the inhomogeneously strained region, and assuming a spheroidal shape of nucleus. He finds that supercooling breaks down when the 'degree of supercooling', defined as

$$\phi = 1 - (H/H_c)^2, \qquad (4.11)$$

is given by

$$\phi = \Delta/z + n, \qquad (4.12)$$

where z is a measure of the thickness of the nucleus, $4\pi n$ is approximately its demagnetizing coefficient, and Δ refers to the bulk metal. We shall see that (4.12) agrees well with experiment.

It is convenient to include at this point a brief discussion of the conditions for formation of normal nuclei in a superconducting body. This problem is not entirely the converse of the one we have just considered, since a normal nucleus can form only at the surface of the superconductor, while a superconducting nucleus can form right inside the normal metal. If a normal region of thickness d and area A is created at the surface (fig. 49), the free energy is increased by

$$A\left\{\left(g_n - g_s - \frac{H^2}{8\pi}\right)d + \frac{\lambda H^2}{8\pi} + \alpha + \alpha_n - \alpha_s\right\},$$

where α_n and α_s are the surface free energies at the free surface of normal and superconducting metal respectively, and λ is the small penetration depth of a magnetic field into a superconductor. This concept of penetration depth will be explained more fully in Chapter v, but here we are concerned only with the fact that the volume of the superconducting phase is effectively reduced by $\lambda H^2/8\pi$, owing to the penetration. Thus such an inclusion can grow only if

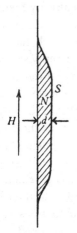

$$(1 - (H/H_c)^2) + (\Delta + \lambda + \beta)/d < 0, \quad (4.13)$$

where $\quad \beta = 8\pi(\alpha_n - \alpha_s)/H_c^2. \quad (4.14)$

We shall see in §5.8 that β is probably very small or zero, so that normal regions can form for $H < H_c$ only if $\Delta + \lambda$ is negative.

Fig. 49. Schematic diagram of flat shaped nucleus of normal phase (area A, thickness d) formed at the surface of a superconductor.

For $H > H_c$, we should find 'superheating' if $\Delta + \lambda$ is positive, i.e. no nucleus of normal state could grow unless its size d was big enough to satisfy (4.13). Rather, as in the case of supercooling, we can suppose that nuclei of normal phase exist in regions where $\Delta + \lambda$ (or some quantity intermediate between Δ and $\Delta + \lambda$ if the nucleus has grown into the interior of the metal) is negative, but actually it is probable that larger nuclei of normal phase occur for other reasons, such as bumps on the surface or regions of frozen-in flux (of the earth's field or of previously applied fields). It is probably for this reason that, as we shall see below, superheating is a much less pronounced phenomenon than supercooling.

4.6.3. *Faber's experiments.* Faber (1949, 1952) has recently verified in a striking way that the type of hysteresis described in §4.6.1 has really the character of a supercooling. A long tin rod (about 20 cm. long) was brought into the 'supercooled' state by lowering the applied field a little, but not too far below H_c. At one place on the rod the field was then lowered further by producing an opposite field in a short coil round the rod, and it was found that for a certain critical value of this 'boosting' field the transition to superconductivity suddenly started at this point of the rod, and then spread in a few seconds along the whole rod. The spreading could be demonstrated by observing ballistic kicks in a galvano-meter connected to search coils round the rod at various points along its length. The growth of a nucleus which then infects the whole rod was thus clearly proved in a manner very similar to that used by Sixtus and Tonks (1931) in the analogous ferro-magnetic problem.

The degree of supercooling ϕ (as defined by (4.11)) varies markedly from point to point along the rod, and can be as large as 0·8. The supercooling of the rod as a whole, however, is governed by the 'weakest' nucleus, which explains why in most experiments only a slight effect is found. If the weakest nuclei are 'cordoned off' by maintaining a field greater than H_c round them, the rest of the rod supercools to an extent governed by the next weakest nucleus and so on.

Equation (4.12) agrees very well with the experimental observa-tions on the temperature variation of ϕ. Thus the curves of ϕ against temperature for different nuclei are all of similar shape, and may be brought into coincidence by suitably displacing the tem-perature axis for each curve,. and then adjusting the scale of ϕ (see fig. 50). The shape of the curves is, moreover, consistent with the temperature variation of Δ suggested by the intermediate state evidence (Table IV). Values of n and z could be deduced for various nuclei by applying (4.12); a typical order of magnitude of n was 0·05, corresponding to a cigar-shaped nucleus with a length 6 times its diameter, while z, the 'thickness', was of order 3×10^{-4} cm.

By applying the local 'boosting' field for only a very short time τ it was possible to show not only that the nuclei were mostly located very close to the surface, but also to estimate their size indepen-dently. It was found, in fact, that only when τ was reduced below about 10^{-4} sec. did ϕ vary at all with τ; this suggests that for shorter times the change of field has not time to penetrate far enough into the surface to go right through the nucleus. From the detailed results it was possible to estimate that the sizes of the nuclei were of order 10^{-4} cm., in agreement with the other estimate.

Once the nucleus is able to grow, its speed of growth is limited mainly by the eddy currents induced in the metal as the flux is ejected from the newly created superconducting regions.* The experiments show that the velocity of growth, v, of the super-conducting phase is proportional to $(\Delta H/H_c)^2$, where ΔH is the difference between H_c and the field in the bulk of the specimen; the

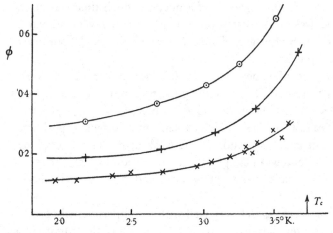

Fig. 50. Temperature variation of ϕ (Faber, 1952). The curves are for three different nuclei; the lowest curve coincides with the behaviour of the specimen as a whole, i.e. shows the variation of ϕ when no 'boosting' coils are used.

constant of proportionality depends in a well-defined way on temperature (see fig. 51), but is independent of the diameter of the rod. Faber is able to account for these features by a mechanism which involves the growth of the nucleus as a filament along the surface, which rapidly spreads sideways over the surface and less rapidly into the interior. The temperature variation, shown in fig. 51, of the constant of proportionality between $v^{\frac{1}{2}}$ and $\Delta H/H_c$, which involves Δ and λ and their temperature variations, is in

* The effect of eddy currents in limiting the rate of movement of a phase boundary has been discussed by Pippard (1950 d) and more recently by Lifshitz (1950). Other experiments bearing on the kinetics of the superconducting transition are those of Alekseyevsky (1945 b, 1948 a), Lasarew, Galkin and Chotkewitsch (1947), Galkin and Lasarew (1948), Galkin, Kan and Lasarew (1950), Galkin, Lasarew and Bezuglyi (1950), Galkin and Bezuglyi (1950), Serin, Reynolds, Feldmeier and Garfunkel (1951), Serin and Reynolds (1952).

reasonable accord with the data on these parameters. The constant itself should depend also on the conductivity of the normal metal, but the form of this dependence has not yet been verified experimentally.* A particularly encouraging feature of this theory is that it explains also the shape of the pulse of e.m.f. developed in the search coil as the transition passes through it (fig. 52); the rapid rise of e.m.f. is associated with the rapid spreading of the superconducting phase over the surface, while the slow decay corresponds to the

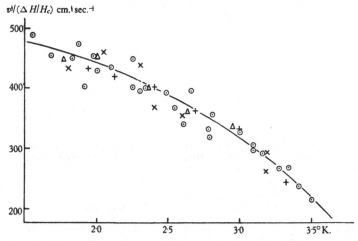

Fig. 51. Temperature variation of the constant of proportionality between $v^{\frac{1}{2}}$ and $\Delta H/H_c$ for various specimen diameters (Faber, to be published).

slower ejection of flux from the interior. Detailed examination of pulse shapes such as that of fig. 52 gives quantitative support for the theory.

It has usually been taken for granted that the converse phenomenon of superheating does not occur, but by very careful control of the field, Faber has been able to demonstrate a slight superheating to fields 1 or 2 % above the true value of H_c. The phenomenon is difficult to observe because of the difficulty of removing traces of normal regions in the specimen below H_c, which, as explained in §4.6.2, act as large nuclei, and permit only a small

* Recent experiments (Faber, to be published) in which the conductivity was varied by alloying have shown that the relation is more complicated than was at first thought; it is probable that mean free path effects of the kind discussed in § 6.3.3 are involved, owing to the small thickness of the superconducting sheath.

degree of superheating. Pippard (1950c) has also demonstrated the existence of superheating by measuring the radio-frequency impedance of a cylindrical wire in transverse magnetic fields; he found that the sudden change of impedance associated with entry into the intermediate state (the analogue of the 'horn' in the magnetization curves (see § 4·5)) occurred at values of ρ which were much higher than those found by Désirant and Shoenberg (1948a). It is probable that these high values of ρ were due to superheating, but it is not clear why such a superheating was not found in the magnetization curves.

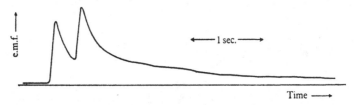

Fig. 52. Pulses of e.m.f. produced in two search coils 10 cm. apart and connected in series as the normal-superconducting transition is propagated through them; the speed of propagation is about 25 cm./sec. (Faber, to be published).

4.7. Destruction of superconductivity by a current

We have already pointed out that when the magnetic field of the current at the surface of a superconductor exceeds H_c the resistance reappears. In the case of a ring, the current then merely adjusts itself to keep the field just critical, and the resistance again disappears, but if the current is maintained constant (by an external source of e.m.f.) the resistance remains. We shall now discuss the changes that take place when superconductivity is destroyed by a current in this way, and for simplicity we shall consider a long cylindrical wire (radius a) with a current i flowing along it.

As soon as the current exceeds the value $\frac{1}{2}aH_c$, the field at the surface becomes greater than H_c and evidently superconductivity will begin to disappear. If we suppose, as at first sight seems plausible, that the transition to the normal state takes place by superconductivity retreating to an inner core of the wire, we get immediately into difficulties somewhat similar to those discussed in § 2.5 in connexion with the transition of an ellipsoid from super- to normal conductivity under the influence of an external magnetic field. In fact, if superconductivity were to retreat to a core of radius smaller than that of the wire, all the current would move into this superconducting core, and consequently produce a field even greater than that originally at the surface of the wire, at the boundary of the core with the normal region surrounding it. The retreat of superconductivity would in fact have to continue until the whole wire was in the normal state; this, however, is obviously impossible, since then the current would be uniformly distributed over the cross-section of the wire, and the field would be less than H_c over the greater part of the cross-section (up to a radius $\frac{1}{2}a^2H_c/i$), so that this part could not be in the normal state. This paradox appears even more strikingly if we consider what happens when a wire carrying a current i is cooled below the transition temperature; since in the normal state the magnetic field of the current is zero on the axis and increases as we move away from the axis, superconductivity should appear first of all on the axis of the wire. If it were to do so, however, all the current would immediately move into the newly created superconducting core, and produce a very large field at the boundary of the core, thus destroying its super-conductivity once again.

Just as in the case of an ellipsoid in an external field, the paradox shows that the transition must take place in a more complicated manner. F. London (1937) and also Landau* have shown that since the core cannot be either superconducting or normal, it must be in the intermediate state, with $H = H_c$. It is easy to see then how the resistance will return as the current i is increased. Suppose that $i > \frac{1}{2}aH_c$, and let x be the total current inside a cylindrical surface of radius r. Everywhere inside the core the field has to be H_c, so that

$$2x/r = H_c, \qquad (4.15)$$

and, in particular, if r_0 is the radius of the core, and x_0 the current carried by it, then

$$2x_0/r_0 = H_c. \qquad (4.16)$$

The current density J is given by $\dfrac{1}{2\pi r}\dfrac{dx}{dr}$, which from (4.15) becomes

$$J = H_c/4\pi r = x/2\pi r^2, \qquad (4.17)$$

---

* Private communication in 1937; Landau has discussed also the destruction of superconductivity by a current in a hollow cylinder (see 1st edition of this monograph, p. 59).

so that at the boundary of the core

$$J_0 = x_0/2\pi r_0^2.$$

Now at the boundary between a normal and an intermediate region the two states pass smoothly into one another, so the current density must be continuous across the boundary. In the normal region outside the core, the current density is constant and equal to $(i - x_0)/\pi(a^2 - r_0^2)$, so that, equating this to J_0, we find

$$2r_0^2\left(\frac{i}{x_0} - 1\right)\Big/(a^2 - r_0^2) = 1. \tag{4.18}$$

Putting $r_0/a = \rho$ and $2i/aH_c = \mu$ (μ is then the ratio of the field at the surface of the wire to the critical field, so that $\mu > 1$), the two conditions (4.16) and (4.18) become

$$x_0/i = \rho/\mu = 2\rho^2/(1 + \rho^2), \tag{4.19}$$

so that

$$1 + \rho^2 - 2\mu\rho = 0, \tag{4.20}*$$

which gives

$$\rho = \mu - \sqrt{(\mu^2 - 1)} \tag{4.21}$$

(the other sign for the square root would make $\rho > 1$).

If the full normal resistance of the wire is R_0, and the resistance for current i is R, the resistance of the normal region surrounding the core will be $R_0 a^2/(a^2 - r_0^2)$ or $R_0/(1 - \rho^2)$; thus in order that the electric field should be constant over the cross-section of the wire,

$$R_0(i - x_0)/(1 - \rho^2) = Ri. \tag{4.22}$$

Substituting (4.19) and (4.21), we find

$$R/R_0 = \tfrac{1}{2}\{1 + \sqrt{(1 - 1/\mu^2)}\}. \tag{4.23}$$

Thus the resistance should rise discontinuously to half its full value as soon as $i = \tfrac{1}{2}aH_c$, ($\mu = 1$), and then continue to rise with further increase of current, reaching its full value only asymptotically. In fig. 53 we show how the resistance should be restored by a current at constant temperature, according to (4.23). Experimentally, this

---

* We may note that (4.20), deduced from the condition that the current density should be continuous across the core boundary, can also be deduced from the condition that the Joule heat developed by the current in the whole wire should be a minimum. The Joule heat W, developed per unit time, is Ri^2, or using (4.16) and (4.22)

$$W = R_0 i \frac{(i - x_0)}{1 - \rho^2} = R_0 i^2 \frac{(1 - \rho/\mu)}{1 - \rho^2}$$

and it can easily be seen by differentiating with respect to ρ that the condition for a minimum is indeed just (4.20).

question is difficult to investigate, on account of the large currents required, since, as soon as any resistance is restored, considerable development of Joule heat occurs, which makes it very difficult to maintain the temperature constant. Schubnikow and Alekseyevsky (1936) overcame this difficulty by the ingenious method of immersing the specimen in liquid helium below the λ-point (helium II), which on account of its enormous effective heat conductivity is able to carry away the Joule heat sufficiently rapidly to prevent the temperature from rising. It is interesting to mention that if the same experiment is carried out above the λ-point, as was tried in the original Leiden experiments on destruction of superconductivity

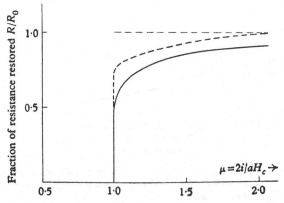

Fig. 53. Restoration of resistance by a current at constant temperature: full curve, theoretical (equation 4.23); broken curve, experimental for tin (Schubnikow and Alekseyevsky, 1936).

by a current, the wire immediately becomes isolated from the liquid by a layer of gas, and the Joule heat is often sufficient to melt the wire, this being accompanied by an explosive boiling of the liquid helium. The experiments of Schubnikow and Alekseyevsky (the broken curve in fig. 53) showed that with a monocrystalline wire of pure tin there was indeed a discontinuous restoration of resistance at exactly the current strength predicted by Silsbee's hypothesis, but the discontinuous rise was up to $R/R_0 \sim 0.8$, instead of to 0.5 as predicted by the theory. This discontinuous rise was followed by a slower rise, but not as slow as predicted by the theory, and R/R_0 reached unity for a current about twice the critical current, in contrast to the asymptotic approach to unity predicted by the theory.

More recently Scott (1948) has carried out similar experiments on indium and has shown that the ratio R/R_0 found experimentally varies with the wire diameter (fig. 54), getting very gradually less as the wire gets thicker. This suggests that the disagreement between theory and experiment has the same origin as the '0·58 effect' discussed in § 4.4, and is associated with the detailed structure of the intermediate state. As already mentioned, however, Silsbee's hypothesis still holds accurately, even though size effects influence the nature of the restoration of the resistance; for a discussion of this point see Pippard (1950 d).

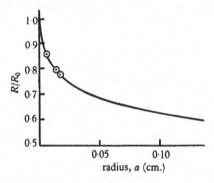

Fig. 54. Variation of R/R_0 (fraction of resistance restored discontinuously by a current) with wire radius, a (Scott, 1948). The points are based on experiments with indium wires, the curve is based on an empirical formula:

$$\log (2R/R_0 - 1) = -(2a)^{\frac{1}{2}},$$

designed to fit the points, to extrapolate to 0·5 for infinite a, and to 1·0 for zero a.

Experiments have also been made on the destruction of superconductivity by a current in the presence of a magnetic field parallel to the current (Alekseyevsky, 1938 a), and, as we should expect, the critical current decreased with increase of the applied magnetic field, vanishing of course when the applied field was equal to H_c. The amount of the discontinuous rise at first fell slightly below 0·8, but for greater fields increased above 0·8, approaching the value unity as the field approached H_c. The presence of an applied magnetic field greatly complicates the problem of determining theoretically how the resistance should be restored, since the field distribution can no longer be considered in only two dimensions. Destruction of superconductivity by a strong current in the presence of a transverse magnetic field has not yet been investigated experimentally. The theoretical problem is in this case further complicated by the absence of circular symmetry in the field distribution, and has also not yet been investigated.

The result (4.23) shows also how the resistance of the wire should disappear when it is cooled with a constant current i flowing in it; in this case the parameter μ varies in virtue of the variation of H_c with temperature. Above the normal transition temperature T_c, H_c is zero and μ infinite, so that $R = R_0$; as soon as the temperature is reduced below T_c, however, μ becomes finite, and decreases with further reduction of temperature, so that the resistance begins to

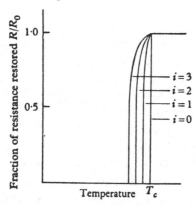

Fig. 55. Influence of current strength on restoration of resistance (London, 1937). The curves are theoretical, based on equation (4.23); the units of i are arbitrary.

drop in accordance with (4.23). This continues, however, only until $\mu = 1$, i.e. until the critical field has reached the value $2i/a$, when the resistance drops abruptly to zero from half its full value and the wire becomes completely superconducting. During the cooling process, the intermediate core appears at temperature T_c, and then grows until, just before the sudden drop of resistance occurs, it occupies the whole of the wire; for any further reduction of temperature, this arrangement becomes unstable, and is replaced by the superconducting state throughout the whole wire.

The variation of resistance with temperature for different currents, as deduced from these considerations, is shown in fig. 55 (for the purpose of this diagram it has been assumed that H_c varies linearly with $(T_c - T)$, which is true for sufficiently small $(T_c - T)$). It can now be understood more precisely why in § 1.1 a vanishingly small measuring current was mentioned as one of the conditions for a completely abrupt appearance or disappearance of resistance. As illustrated in fig. 3b, transition curves similar to those shown in fig. 55 have in fact been observed experimentally by de Haas and Voogd (1931c), but the currents used were rather too small to allow of more than a qualitative confirmation of the theory.

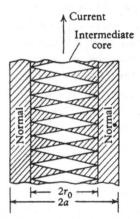

Fig. 56. Schematic diagram of structure of a cylindrical wire in which resistance has been partially restored by a current.

We have said nothing, so far, about the structure of the metal in the intermediate core, which will evidently be different from that discussed in §4.2, since here, on account of the current, the lines of force are circular instead of straight. From the expression (4.17) for the current density inside the core, we see that the intermediate state must consist of a mixture of superconducting and normal regions arranged somewhat on the lines of fig. 56. Thus, since the current density increases towards the axis as $1/r$, while the electric field is constant, the resistance of a filament at distance r must increase proportionally with r, and so the thickness of the laminae must also vary linearly with r, the normal laminae growing thicker away from the axis, until they fill up the whole length of the wire at the core boundary, the superconducting laminae growing thicker towards the axis. On the axis itself the metal would, according to these considerations, be completely superconducting, but it can easily be seen that a very thin filament round the axis would have to carry only a vanishingly small current, so that there is no contradiction. Just as for an ellipsoid in an external field, the scale of the structure, i.e. the number of laminae per unit length of the wire, depends on the ratio of the radius of the core to the characteristic parameter Δ, but no detailed theory has been worked out. To reduce the curvature of the lines of current flow (which must enter the superconducting regions normally) as much as possible, there should be many laminae, but to reduce the surface energy there should be few, and it is clear that a maximum-minimum problem is involved. The problem is, however, complicated by the fact that it is almost certainly not permissible to use the concept of surface energy where two laminae approach very close together (i.e. near the axis) and moreover if the normal laminae are thin, mean-free path effects (see §6.3.3) may be relevant.

CHAPTER V

THE DEPTH OF PENETRATION OF
A MAGNETIC FIELD INTO
A SUPERCONDUCTOR

5.1. Introduction

We have seen that the characteristic property of a superconductor is that $B = 0$, and consequently that any current flowing must be superficial. It is, however, obvious that the current cannot be entirely superficial, i.e. that there must be a certain depth of penetration of the current, which will also be the depth of penetration of the magnetic field into the superconductor. Since the $B = 0$ description is adequate for specimens of macroscopic size (~ 1 cm.), it is evident that this penetration depth must be very small compared with 1 cm., and in this chapter we shall give an account of the various methods by which this depth (which we shall always denote by λ) has been studied. Considerable progress has been made in this part of the subject since the first edition of this monograph appeared, and it is now possible to answer fairly definitely many questions which were then quite open.

In our treatment we shall depart from the historical order of developments in two respects. In fact, the theory of F. and H. London (1935 a) predicted the detailed form of many of the results to be discussed long before any of the experiments were made; it will, however, be convenient to deal with this theory in Chapter VI, and in the present chapter we shall merely make forward references where appropriate. Again, it was the resistance measurements of Shalnikov (1940 a) and of Appleyard, Bristow, H. London and Misener (1939) on thin metal films which in fact provided the first indication that penetration effects were important for thin specimens in producing very high critical fields.* It has, however, since turned out that the interpretation of their measurements is less straightforward than was originally thought, and so they will be described only in §5.7 after the more direct magnetic evidence has

---

* Preliminary notes on this effect appeared at about the same time (Shalnikov, 1938; Appleyard and Misener, 1938).

been presented. The first direct evidence came from the magnetic properties of mercury colloids, i.e. small spheres (§5.2.1); these experiments showed not only that a magnetic field penetrates appreciably, but that the penetration depth is temperature-dependent. Similar experiments on thin wires (§5.2.2) and films (§5.2.3) have since been made. At first sight it might seem that the effect of the penetration depth could be studied only by using a specimen of a size comparable to λ, but as Casimir (1940) first pointed out, the fact that λ varies with temperature permits some very useful information about it to be obtained from accurate measurements on macroscopic specimens; the realization of his idea, both at low frequencies by Laurmann and Shoenberg (1947, 1949), and at very high frequencies by Pippard (1947c), is described in §5.3. The possible dependence of λ on direction of the current with respect to the crystal axes of a single crystal is discussed in §5.4 and the possible dependence on field strength in §5.5. Another important question is the law according to which the field decays on entering the specimen; the rather meagre evidence on this question is presented in §5.6. After a discussion of the significance of the area of the magnetization curve of a small specimen and of the critical field measurements on thin films, which involves a generalization of the thermodynamic theory of Chapter III, the chapter concludes with a brief account of the problem of critical currents in thin specimens (on which there is as yet practically no information).

5.2. Magnetic properties of small specimens

For the measurement of magnetic properties the difficulty of reducing the dimensions of a specimen is that the volume, and hence the magnetic moment to be measured, goes down too. From this point of view it would seem advantageous to reduce only *one* dimension (i.e. to work with thin plates) rather than reduce two dimensions (thin wires) or three (small spheres), but the more rapid reduction in volume of wires and spheres can be compensated by using a compound specimen ideally consisting of a large number of replicas. In using this replica method, care must be taken that the units are sufficiently far apart for their mutual interactions to be reasonably small, and evidently the units must be as nearly as possible identical if we are to obtain results characteristic of a certain size, rather than an average. In fact, the first investigation of this kind was done with a large number of spheres in a colloid (Shoenberg, 1940a), and it is only fairly recently that wires (Désirant and Shoenberg, 1948b) and plates (Lock, 1951) have been investigated.

Before describing the results in detail it will be convenient first to define the concept of penetration depth a little more precisely, and to indicate how we should expect λ to enter into the magnetic properties. If the field at depth x inside the surface of an infinitely thick specimen has fallen from its value H_0 at the surface to a value $H(x)$, then we define the penetration depth as

$$\lambda = \frac{1}{H_0} \int_0^\infty H(x)\,dx. \tag{5.1}$$

Thus, if the field fell off exponentially (as, indeed, is predicted by the Londons' theory—see § 6.2.1), then λ would determine the rate of decay, i.e. the exponential law would be

$$H = H_0 e^{-x/\lambda}. \tag{5.2}$$

Again, if the field were to remain constant to a certain depth and then suddenly vanish, this depth would be just λ (this variation is of course unplausible, but is sometimes useful as a crude assumption for qualitative arguments). If the specimen has finite dimensions it is to be expected that λ will still determine the field distribution inside, but exactly how will in general depend upon the fundamental law governing the variation of field. Thus the Londons' theory, for instance, predicts that the variation of H is governed by the differential equation $\nabla^2 H = H/\lambda^2;$ (5.3)

and the variation in finite specimens is then calculable in simple cases (see §6.2.1 and Appendix II). We can, however, make some predictions about the field variation for two extreme cases, independently of any specific assumption such as (5.3). These are for specimens which are either *large* or *small* compared with λ. For large specimens it is evident that the form of the field decay at one boundary of the specimen will not be appreciably changed by the existence of a second boundary at some value of x which is much larger than λ. We can, therefore, safely assume that $H(x)$ is the same as for a semi-infinite body; the distribution of field is shown schematically in fig. 57a. For small specimens it is reasonable to suppose that the field penetrates the specimen almost completely; the distribution of field across the specimen (taken for concreteness as a plate parallel to the field) must have a form somewhat as shown in fig. 57b. Since there must be a minimum at the centre, it is evident that such a distribution can be represented by a parabola if the drop between the surface and the centre is sufficiently small.

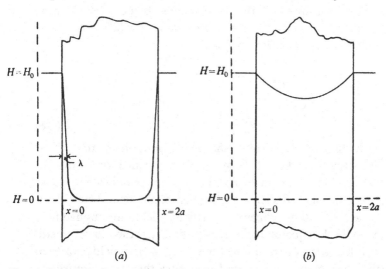

Fig. 57. Schematic diagram showing field distribution in a superconducting plate placed in a magnetic field H_0 parallel to its surface; (a) for $a \gg \lambda$; (b) for $a \ll \lambda$.

Now provided that there is some fundamental differential equation governing the field decay, the form of the parabola must clearly involve the same λ which is defined by (5.1) for a macroscopic specimen, and we can write

$$H = H_0[1 - \gamma(a^2 - (x-a)^2)/\lambda^2], \qquad (5.4)$$

where γ is a numerical constant depending on the nature of the theory of the penetration, and involving the geometry of the specimen. This equation really states nothing else than the plausible assumption that the scale of the decay curve is determined by λ.

From these considerations the magnetic susceptibility is easily deduced. For *large* specimens the effect of penetration of the field is entirely equivalent to the removal of a layer of thickness λ from the specimen. If κ_0 is the susceptibility (defined as I/H_e) for a specimen with no penetration, i.e. $-1/4\pi$ for a wire or plate parallel to the field and $-3/8\pi$ for a sphere (we shall refer to this as a 'bulk' susceptibility), then for a finite specimen with $a \gg \lambda$

$$\kappa/\kappa_0 = 1 - \lambda/a \quad \text{for a plate of thickness } 2a, \quad (5.5)$$

$$\kappa/\kappa_0 = 1 - 2\lambda/a \quad \text{for a wire of radius } a, \quad (5.6)$$

$$\kappa/\kappa_0 = 1 - 3\lambda/a \quad \text{for a sphere of radius } a. \quad (5.7)$$

For a very small specimen the answer is less obvious. Consider the plate illustrated in fig. 57b; the *average* value of B (B can of course no longer be assumed uniform) is

$$\bar{B} = \frac{1}{a}\int_0^a H\,dx = H_0(1 - 2\gamma a^2/3\lambda^2),$$

so that $\qquad \kappa = (\bar{B} - H_0)/4\pi H_0 = -\gamma a^2/6\pi\lambda^2,$

or $\qquad\qquad\qquad \kappa/\kappa_0 = \alpha a^2/\lambda^2, \qquad\qquad\qquad (5.8)$

where α is a new numerical coefficient whose value for the plate is $\frac{2}{3}\gamma$; it is easy to see that the same result is obtained for a wire, but with $\alpha = \frac{5}{6}\gamma$.* For a sphere the argument is more complicated because (5.4) is no longer sufficient specification of the field distribution, since the field will no longer be exactly parallel to the applied field H_e except in the equatorial plane. Equation (5.4) can, however, still be regarded as the distribution of the component of field *parallel* to H_e (the other components are irrelevant, since they do not contribute to the magnetic moment), provided γ is now regarded as a function of y, the depth of the plane concerned below the centre of the sphere. The full calculation of κ/κ_0 has little point since γ is in any case an unspecified parameter, but it is evident that the final result (5.8) must be obtained, with some suitable value of α.

The importance of these considerations is in showing that experiments with specimens large compared with λ in which the variation of κ/κ_0 with a or λ is studied (we shall see below that λ can be varied by changing the temperature), are useless for investigating the truth of any particular penetration law such as §5.3. Experiments on very *small* specimens can only be used for this purpose if the absolute values of a and λ are known, for then the coefficient α can be determined and compared with the value from a particular

* Note that γ has not the same value for a plate and a wire.

penetration law (thus, as we shall see in Appendix II, the values of α predicted by the Londons' law (5.3) are $\frac{1}{3}$ for a plate, $\frac{1}{8}$ for a wire and $\frac{1}{15}$ for a sphere).

5.2.1. *Spheres.* The first clear indication of penetration effects in magnetic measurements came from experiments by the author on mercury colloids using the method of § 2.11.3 (Shoenberg, 1939, 1940 a). We shall deal first with the results on one particular preparation whose particle size was somewhere between 10^{-6} and 10^{-5} cm., and proved to be small compared with λ. The difficulty about all the colloid measurements was that the particle size was in fact known only to an order of magnitude, and, moreover, the 'ideal' condition of identical spheres mentioned in § 5.1 was not fulfilled in so far as there was a big spread of particle sizes. Fortunately, this last feature is not disastrous in the application of (5.8), since if a varies from sphere to sphere, we still have that for the whole assembly of spheres,

$$\kappa/\kappa_0 = \alpha a'^2/\lambda^2, \qquad (5.9)$$

where a' is a suitable average value of a (defined by $a'^2 = \overline{a^5}/\overline{a^3}$), provided all the particles satisfy the condition $a \ll \lambda$ adequately.

The magnetization curves for this particular specimen are shown in fig. 58 (see also fig. 73), and the values of κ/κ_0 are plotted as a function of temperature in fig. 59. The fact that κ/κ_0 is less than 10^{-2} suggests that (5.9) may be safely applied. The variation of κ/κ_0 with temperature can presumably be due only to a variation of λ with temperature, and the form of this variation can be deduced from the results by applying (5.9), though since a' is unknown apart from its order of magnitude, only relative values of λ can be deduced. The variation of λ/λ_0 with T is shown in fig. 60 (λ_0 is the value of λ at $0°$ K.), and fig. 61 shows that the data are remarkably consistent with the law

$$(\lambda/\lambda_0)^2 = 1 \bigg/ \left(1 - \left(\frac{T}{T_c}\right)^4\right), \qquad (5.10)$$

which as we shall see in § 6.3.2 has some theoretical basis. It should be noticed that close to the transition temperature this reduces to

$$\lambda \propto (T_c - T)^{-\frac{1}{2}}. \qquad (5.11)$$

It is reassuring that the value of T_c which was found experimentally for the mercury colloid was $4\cdot15°$ K.—very close to the accepted transition temperature of mercury; this supports our interpretation that the reduction of κ is due to penetration effects rather than to any secondary effect associated with the reduction of size.

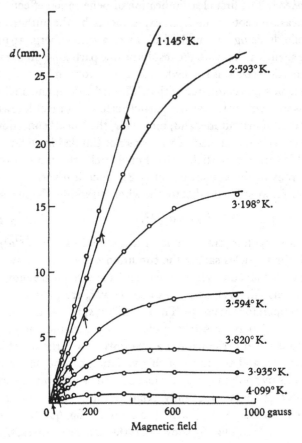

Fig. 58. Magnetization curves of mercury colloid (Shoenberg, 1940a). The galvanometer deflexion, d, is proportional to the magnetization. The arrows indicate the values of H_e for bulk mercury at the various temperatures.

If we assume the value of $\lambda_0 = 4 \times 10^{-6}$ cm. indicated by later measurements (see §5.3.1), and assume $\alpha = \frac{1}{15}$ we can deduce that $a' = 1 \cdot 5 \times 10^{-6}$ cm.; this is rather lower than was expected on the basis of some optical measurements, which indicated 5×10^{-6} cm., but the reliability of this estimate was uncertain, so that the discrepancy can hardly be regarded as proving that the Londons' value of α is wrong.* It would indeed be very valuable to have more measurements on small spheres in which a' is reliably known.

It is convenient to mention here two features of the magnetization curves of small spheres; namely, the shape of the magnetization curve, and hysteresis probably associated with supercooling. For the very small spheres of the colloid just discussed it is difficult to make any certain deduction about the shape of the magnetization curve owing to the variety of sizes present (see § 5.5), but it proved possible to make a much more homogeneous preparation with rather

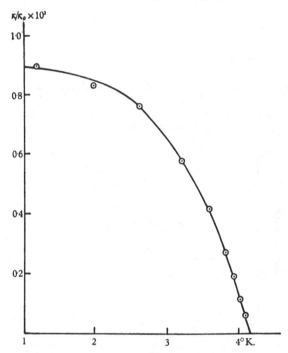

Fig. 59. Temperature variation of κ/κ_0 for the mercury colloid of fig. 58 (Shoenberg, 1940a).

larger particle size (radius of order 10^{-4} cm.). The magnetization curves of this new preparation approached closely to right-angled triangles, suggesting that the intermediate state region (see fig. 11 b)

* Recently Lock (unpublished) has examined the particles in this colloid by means of the electron microscope, and found a value of a' of order 2×10^{-6} cm., which is reasonably consistent with the value deduced from the magnetic measurements assuming $\alpha = \frac{1}{15}$.

was practically absent. This is just what we should expect, since the intermediate state occurs only because in the superconducting state the field is excluded from a macroscopic sphere and so becomes higher at some parts of the surface than others; for a small enough

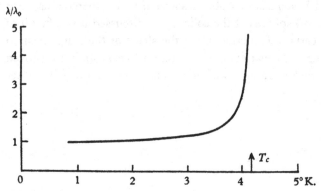

Fig. 60. Temperature variation of λ/λ_0 as deduced from the data of fig. 59 (Shoenberg, 1940 a).

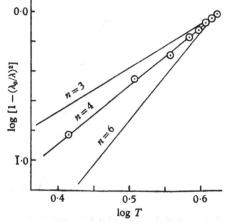

Fig. 61. Plot of $\log (1-(\lambda_0/\lambda)^2)$ against $\log T$ to illustrate validity of (5.10). The lines marked $n = 3, 4, 6$ are plotted on the assumption of 3rd, 4th and 6th powers in (5.10) (Daunt, Miller, Pippard and Shoenberg, 1948).

sphere, however, penetration of the field becomes so appreciable that this variation of field over the surface is much reduced, and the intermediate state region is correspondingly contracted. For still smaller spheres the magnetization curve may of course depart again from the right-angled triangle shape, if the penetration depth is not independent of field strength; this possibility is discussed in §§ 5.5 and 6.3.5.

The second feature to be noted here is that there was practically no hysteresis in the magnetization curves of the very small sphere colloid (figs. 58 and 71), i.e. there appeared to be no appreciable supercooling. For the medium-sized spheres (radius $\sim 10^{-4}$ cm.), however, there was a very considerable 'supercooling' (see § 4.6.1), which became more marked as the temperature approached T_c; the greatest supercooling was to something like one-quarter or one-half of the critical field. These features find a qualitative explanation (Pippard, 1952) in terms of the two-fluid theory as modified to take account of the field dependence of penetration depth, and an outline of this explanation will be found in § 6.3.5.

Other features of the magnetization curves of the very small spheres will be discussed in § 5.5 (possible field dependence of λ), §§ 5.7.3 and 5.8 (critical fields) and § 6.3.5 (theoretical discussion).

5.2.2. *Thin wires.* Désirant and Shoenberg (1948 b) measured the magnetization curves of thin mercury wires (in glass capillaries); sufficient volume of mercury to make measurements by the method of § 2.11.3 was obtained by stacking together as many as 200 filled pieces of capillary, each about 1 cm. long. The thinnest wires used were of radius about 5×10^{-4} cm., and it is unlikely that it will prove possible to work with much thinner wires. Since λ is of order 10^{-6} to 10^{-5} cm., it is clear that (5.6) applies and κ/κ_0 would not be expected to differ from unity by more than a few per cent. In fact the method of measurement was not accurate enough *absolutely* to permit measurement of such small differences between κ and κ_0,[*] but it could show the *changes* of κ/κ_0 due to the temperature variation of λ. In fact we have from (5.6) that

$$\frac{\Delta\kappa}{\kappa_0} = \frac{\kappa(T_0) - \kappa(T)}{\kappa_0} = \frac{2[\lambda(T) - \lambda(T_0)]}{a}, \qquad (5.12)$$

where T_0 is some convenient low temperature (usually about 2° K. in these experiments), so that from accurate measurements of $\Delta\kappa/\kappa_0$ (which amounted to a few per cent and could be measured with a precision of order 5 or 10 %), $\lambda(T) - \lambda(T_0)$ could be deduced in absolute measure. If now $\lambda(T)/\lambda(T_0)$ is taken from the colloid

[*] In any case the specimen volume was never known to better than a few per cent.

measurements (or from (5.10) which represents these measurements quite accurately), the absolute value of $\lambda(T_0)$ (and hence of λ_0) is immediately obtained.

This method gave $\lambda_0 = 7 \cdot 6 \times 10^{-6}$ cm. for mercury, but as we shall see in § 5.3, this value is probably nearly twice the true answer. The discrepancy is probably due to an insufficiently sharp transition of the compound specimen. Indeed, any spread of transition temperature could cause an exaggerated estimate of λ_0, since evidently if the specimen is incompletely superconducting below its 'mean' transition temperature (which is all that can be determined), the gradual diminution of the normal fraction as the temperature is lowered will, in this experiment, be indistinguishable from a diminution of penetration depth. We shall see in § 5.3.1 that in fact mercury frozen in glass can have rather smeared transitions of this sort, and also that frozen-in remnants of the flux of the earth's magnetic field (which was not compensated in these experiments) can produce a small temperature-dependent residuum of non-superconducting metal. Moreover, since the specimen consists of many separate pieces of mercury wire, it would in any case be surprising if the transition were not a little spread due to individual wires being in different states of stress.

5.2.3. *Thin films.* Quite recently Lock (1951) has completed measurements of the magnetization curves of films of tin, indium and lead deposited by evaporation on mica. While with mechanically rolled films there is great difficulty in producing films as thin as 10^{-4} cm., with this method there is no difficulty in producing films only 10^{-5} cm. thick. In fact the limitation is not the difficulty of producing thinner films, but merely the difficulty of getting together in the space available enough material for the measurements. The specimens were stacks of as many as 300 small pieces cut from a large sheet of mica on to which a film was evaporated through a grid of circular holes (about 6 mm. diameter).

It is well known that the electrical properties of evaporated thin films depend very markedly on the cleanliness of the conditions of deposition; in particular, reproducible results are obtained only if the vacuum is at least of order 10^{-7} mm. and if the films are measured *in situ* (see Appleyard and Bristow, 1939). Such stringent conditions were not possible in Lock's experiments, but it was hoped that since magnetization is a bulk property it would not be too much affected by any breaking up of the film such as would be disastrous for electrical measurements. This hope seems to have been justified, for the measurements have proved to be reasonably consistent not only with each other but also, for tin, with data obtained by entirely different methods.

One of the main inaccuracies of the experiment arises from the fact that the susceptibility of a plate is much greater if the field is transverse rather than parallel to the plate, so that very accurate parallelism (better than $1°$) is necessary to avoid an exaggerated estimate of κ. Special precautions were taken to achieve this, and a further empirical correction procedure was applied to allow for any slight non-parallelism of the films within the stack. The transition temperature of most of the films (particularly for tin) differed considerably—by about $0.2°$ K.—from the transition temperature of bulk metal. This was shown to be associated with stresses produced in the film by the different thermal expansions of the metal and the base. Thus, if distrene foil which has a thermal expansion greater than that of tin was used in place of mica which has an expansion less than that of tin, the transition temperature was lowered instead of raised. It was, indeed, possible to produce films in which the transition temperature was practically right by depositing the tin on an aluminium base, but since the aluminium could not be made as flat as mica, the results with such films were less satisfactory in other respects than those using mica.* The increased transition temperature was taken into account by always using for T_c in the theoretical formulae the actual value for the particular specimen; the consistency of the values of λ_0 found suggests that λ_0 is not very sensitive to variations of T_c.

For the thicker films, equation (5.5) applies, and combining it with (5.10), a value of λ_0 could be deduced from the temperature variation of κ/κ_0 for each specimen. For the thinner films (5.5) must be replaced by (5.17) as explained in §5.6, and again by a suitable analysis a value of λ_0 can be deduced from the temperature variation of κ/κ_0. The average values of λ_0 obtained in this way are shown in

---

* A 'sandwich' technique in which the tin was evaporated on to an aluminium film previously evaporated on mica, produced less increase of T_c than direct deposition on the mica, but the improvement was too slight to justify the trouble involved in using this technique.

Table V. The value of λ_0 for tin is in excellent agreement with values obtained by other methods; those for lead and indium are the only determinations. Other features of the magnetization curves of the plates will be discussed later; non-linearity in § 5.5, validity of the penetration law in § 5.6 and critical fields in § 5.7.

TABLE V. *Values of λ_0 by various methods* (cm. $\times 10^6$)

Metal	Lock (1951)	Laurmann and Shoenberg (1949)		Pippard (1947c)
In	6·4	—		—
Sn	5·0	5·2		5·0 (polycrystal)
		‖	⊥	
Hg	—	3·8	4·5	—
Pb	3·9	—		—

5.3. Experiments on macroscopic specimens

5.3.1. *The Casimir experiment.* As already mentioned in § 5.1, Casimir (1940) suggested that if λ varied with temperature as strongly as suggested by the colloid experiments, it should be possible to reveal this temperature variation by careful measurements on macroscopic specimens. More specifically, he suggested a method rather like that of § 2.11.6, in which a mutual inductance contained the cylindrical specimen as a closely fitting core. Very precise measurements of the mutual inductance (with a low-frequency a.c. method) could then show the slight changes of field penetration with temperature. If the radius of the specimen is a, the difference between the mutual inductance with the specimen removed (M_0), and present (M) is

$$M_0 - M = k\pi a^2(1 - 2\lambda/a); \qquad (5.13)$$

where k is a constant, which, if required, can be calculated from the coil geometry. Thus

$$M(T) - M(T_0) = 2k\pi a[\lambda(T) - \lambda(T_0)],$$

or
$$\frac{M(T) - M(T_0)}{M_0 - M} = \frac{2[\lambda(T) - \lambda(T_0)]}{a}. \qquad (5.14)$$

In (5.14) we have neglected λ/a compared with 1, since it is only 10^{-4} at most; T_0, as in § 5.2.2, is some convenient low temperature. Just as in the method of § 5.2.2, λ could in principle be determined absolutely from (5.13), but this would involve knowledge of k and

a to 1 part in 10^6, and is in fact quite out of the question. Since, however, M itself can be very precisely measured, it is possible to measure the small changes in it (of order a few tenths of a microhenry for a practical set up) caused by temperature changes and hence from (5.14) to deduce the changes of λ. Combining such differences with the relative values given by (5.10), an absolute value of λ_0 can be found.

Evidently, since the changes in M are small, great care must be taken to avoid (or allow for) any small changes due to subsidiary causes, and Casimir was unlucky in overlooking one such cause which masked the genuine effect. His coils were wound directly on a sealed off fused quartz tube containing the mercury specimen, and as the helium vapour pressure was reduced (to reduce the temperature) this probably caused a slight swelling of the tube and hence a slight increase of M which roughly compensated the decrease caused by the change of penetration depth. Casimir's experiment thus appeared to produce a negative result, and it was only in 1947 that Casimir (in a private discussion with the author) suggested the probable cause explained above. The experiment was repeated by Laurmann and Shoenberg (1947, 1949) using 70 cyc./sec., and has given positive results for tin and mercury.

The most serious difficulty in obtaining clear-cut results proved to be that of obtaining sufficiently sharp transitions from the normal to the superconducting state. One cause of smeared transitions, which had already been pointed out by Casimir, was the freezing-in of the earth's magnetic field in imperfections. Experiment showed that it was mainly the component perpendicular to the specimen axis which was effective, and its effect was to leave normal regions below the transition temperature; as the temperature was lowered these regions shrank, and the penetration of the field thus appeared to change by much more than could be accounted for by the genuine superconducting penetration depth. This effect was eliminated by careful compensation of the earth's magnetic field, but much more difficulty was experienced in dealing with the natural spread of the transition; fortunately, there is a 'tell-tale' indication which enables a clear distinction to be made between the genuine effect and any spurious apparent reduction of penetration due to the disappearance of the last traces of normal state some way below the transition temperature. This 'tell-tale' is the presence of an 'in phase' or loss component M' in the mutual inductance (so that the mutual inductance is $M - jM'$), which is partly due to losses in the circuit but partly to the presence of normally conducting regions in the specimen. Thus if M' varies with temperature we have a clear indication that there are normal regions present and that their size is varying. Indeed, it was found possible to relate the change in M' to that part of the change of M *not* due to change of λ, by experiments in which these effects were predominant (e.g. with the earth's field uncompensated). By means of this relation, it was possible to separate the 'spurious' from the genuine change in M, provided the 'spurious' change due to a spread-out transition was not too large. The best results were obtained with cast single-crystal specimens. For mercury it was found essential to remove the containing vessel in which the rod had been cast before cooling to liquid helium temperatures. Specimens cooled in their containing tubes appeared to stick to the fused quartz surface, and always had spread-out transitions due presumably to the stresses caused by the impossibility of free thermal contraction.

Some typical results illustrating the course of an experiment are shown in figs. 62 and 63 for one of these free mercury rods. In the normal state $M - M_s$ (M_s is the value of M in the superconducting state) and M' are quite comparable (as can be expected from eddy current theory), and the change to superconductivity in which M drops to M_s and M' practically vanishes, takes place in a very narrow temperature range as can be seen from fig. 62. The further course of the experiment is shown in fig. 63, where changes of M are now on a thousand-fold expanded scale. It will be seen that for this specimen M' hardly changes at all below $4 \cdot 15°$ K., and the changes of M can, therefore, with confidence be attributed to changes of λ in accordance with (5.14). For tin the results were similar, though, as can be seen from fig. 64, $\Delta M' = M'(T) - M'(T_0)$ is not quite so negligible as for the mercury specimen. As explained above, a small correction proportional to $\Delta M'$ was subtracted from the experimental values of $\Delta M = M(T) - M(T_0)$ before (5.14) was used to deduce values of $\lambda(T) - \lambda(T_0)$; the corrected curve is marked $\Delta M_{corr.}$ in fig. 64.

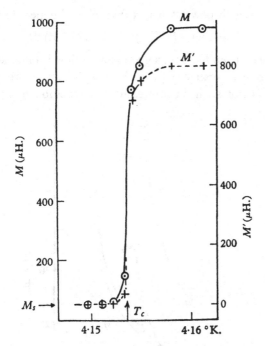

Fig. 62. Variation of M and M' in the transition to superconductivity for a good mercury specimen (Laurmann and Shoenberg, 1949).

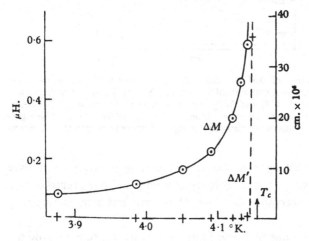

Fig. 63. Variation of ΔM and $\Delta M'$ in the superconducting state for a good mercury specimen (Laurmann and Shoenberg, 1949). The right-hand scale of ordinates translates the changes of M into changes of λ by means of (5.14).

Applying the fourth-power law (5.10) we should expect to find

$$\lambda(T) - \lambda(T_0) = \lambda_0[(1 - (T/T_c)^4)^{-\frac{1}{2}} - (1 - (T_0/T_c)^4)^{-\frac{1}{2}}],$$

and as can be seen from fig. 65 a, a good straight line is obtained if $\lambda(T) - \lambda(T_0)$ is plotted against $(1 - (T/T_c)^4)^{-\frac{1}{2}}$, which we denote by z. Actually, this is not such an impressive confirmation of (5.10)

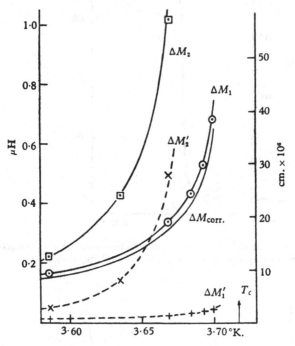

Fig. 64. Variation of ΔM and $\Delta M'$ in the superconducting state for a tin specimen (Laurmann and Shoenberg, 1949). ΔM_1 and $\Delta M_1'$ with the earth's field compensated, ΔM_2 and $\Delta M_2'$ with the earth's horizontal field uncompensated; $\Delta M_{\mathrm{corr.}}$ is the result after correction as explained in the text. The right-hand scale of ordinates translates the changes of M into changes of λ by means of (5.14).

as might at first sight appear, since the range of temperature over which $\lambda(T)$ varies appreciably is rather small; if, however, the truth of (5.10) is accepted the slope of the straight line immediately gives λ_0.

For tin the value of λ_0 was found to be $5 \cdot 0 \times 10^{-6}$ cm. and did not appear to depend on whether the tetragonal axis was parallel or perpendicular to the field. For mercury there seemed to be a slight dependence on crystal orientation, from which it could be deduced (see §5.4) that for current flow perpendicular to the trigonal axis $\lambda_{01} = 4\cdot5 \times 10^{-6}$ cm., while for current flow parallel to the trigonal axis $\lambda_{03} = 3\cdot8 \times 10^{-6}$ cm. The reliability of this evidence for an anisotropic effect is discussed in §5.4. Evidence from this experiment on the possible field dependence of λ is discussed in §5.5.

5.3.2. *Shalnikov and Sharvin's experiment.* A method somewhat similar to that of Casimir has been used by Shalnikov and Sharvin

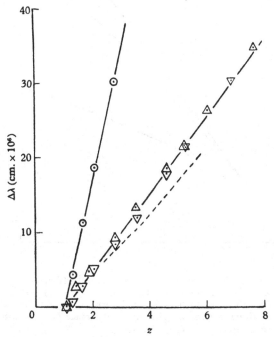

Fig. 65a. Variation of $\Delta\lambda (=\lambda(T)-\lambda(T_0))$ with z ($=(1-(T/T_c)^4)^{-\frac{1}{2}}$) for tin (Laurmann and Shoenberg, 1949). $\triangledown$ as cast, $\odot$ after acid etching, $\triangle$ after electrolytic polishing. The broken curve is for $\Delta\lambda'$ based on results by Pippard (1947c) at 1200 mcyc./sec.

(1948), though so far with little success, owing probably to unsatisfactory specimen conditions. In this experiment the temperature was oscillated instead of the primary current in the mutual inductance. The temperature oscillation was produced by means of a bellows arrangement which oscillated the bath pressure at 4 cyc./sec. The change of λ with temperature now caused a small periodic change of flux in the secondary coil for a fixed current in the primary, thus producing an alternating e.m.f. proportional to $d\lambda/dT$. By integration of the results, and combination with the limiting form of the temperature variation of λ near T_c (according to (5.11)), values of λ were deduced for the single tin specimen studied. The values found were several times higher than those deduced by

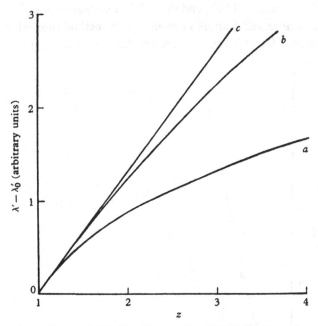

Fig. 65b. Variation of $\lambda'-\lambda_0'$ with z ($=(1-(T/T_c)^4)^{-\frac{1}{2}}$) (Pippard, 1950a). (a) 9400 mcyc./sec. (b) 1200 mcyc./sec. (Pippard, 1947c). (c) 70 cyc./sec. (Laurmann and Shoenberg, 1949). Curves c and b are the straight line and the broken curve respectively of fig. 65a, slightly adjusted to make $T_0=0$, and shown on a somewhat expanded scale of z.

Laurmann and Shoenberg, and it is probable that the discrepancy is due to normal regions persisting below the transition temperature —the same feature that caused so much trouble in the application of the Casimir method.* Since the earth's field was compensated in

* It is just possible that the large values of λ may have been due to a macro-scopic roughness of the specimen; provided the scale of the roughness is not small compared with λ this would cause an increase in the apparent value, in the ratio of the true to the apparent perimeter. Such an effect was, for instance, well marked in an etched specimen studied by Laurmann and Shoenberg, and could be eliminated by electrolytic polishing (see fig. 65a). Since Shalnikov and Sharvin's specimen was mechanically polished it is unlikely that this explanation can account completely for the large observed values of λ.

the Shalnikov and Sharvin experiment, such a smearing of the transitions can have been caused only by specimen conditions, and it is possible that the mechanical polishing which was used in the final preparation of the specimen may have been responsible. The 'tell-tale' of a loss component was not available in this method, so this interpretation can of course be only speculative, and it is to be hoped that the discrepancy will be cleared up by a repetition of this experiment with a greater variety of specimen conditions.

5.3.3. *High frequency reactance method.* In the course of his radio-frequency measurements of resistance Pippard (1947c) noticed that the resonance frequency ν of his specimen changed slightly (by about 1 part in 10^4) when superconductivity was destroyed by a magnetic field, and this could be explained by the difference between the penetration of the radio-frequency field into the wholly superconducting and into the wholly normal metal. If the penetration depth in the normal state (associated with the anomalous skin effect—see Pippard (1947b) and §6.3.3) is δ_n, and that in the superconducting state is λ' (which, as we shall see, is not necessarily the same as the penetration depth λ for a low-frequency or a steady field), then it can be shown that

$$\Delta\nu \propto \delta_n - \lambda'.$$

By a suitable calibration procedure Pippard was able to deduce the constant of proportionality, and from measurements of $\Delta\nu$ to deduce values of $\delta_n - \lambda'$ at various temperatures. Since for tin the electrical resistance is in the 'residual resistance' range, i.e. independent of temperature, it can be safely assumed that δ_n does not vary with temperature, and so values of $\lambda'(T) - \lambda'(T_0)$ could be found by subtraction. The results for a tin specimen at 1200 mcyc./sec. are shown in figs. 65a and 65b, plotted against z ($=(1-(T/T_c)^4)^{-\frac{1}{2}}$), and it can be seen that for low temperatures (i.e. z small) they agree well with the corresponding data for λ, but fall away from the straight line as T_c is approached. The cause for this falling away can be understood qualitatively in terms of the two-fluid model which will be discussed in more detail in §6.3.3, as due to the effect of the radio-frequency field on the 'normal' electrons whose proportion grows as the temperature rises. If this interpretation is sound, λ'_0 and λ_0 are identical, and the slope of the linear part of the plot should give λ_0; the value obtained in this way was $5\cdot0 \times 10^{-6}$ cm., in good agreement with the various other estimates (see Table V) already described (though in fact Pippard's estimate of λ_0 was the first for tin). Pippard has also measured $\delta_n - \lambda'$ for mercury, but here it is probable that δ_n varies with temperature, so no simple interpretation is possible.

More recently Pippard (1950a, b) has used the same method at the higher frequency of 9400 mcyc./sec. to study the variation of penetration depth with crystal orientation and to investigate the field dependence of penetration depth. These experiments will be discussed in §§ 5.4 and 5.5, but we may mention here that the qualitative interpretation of the difference between λ' and λ as due to the presence of normal electrons is supported by the fact that the difference is much more marked at 9400 mcyc./sec. (see fig. 65b), which is, of course, to be expected, since the effect of normal electrons must increase with increasing frequency.

This technique, which in principle amounts to measuring the changes of high-frequency self-inductance due to changing penetration, has two important advantages over the low-frequency method. The first is that the currents in Pippard's specimens flow up and down the length of the specimen, so that in the study of crystal orientation effects the current has a unique direction with respect to the crystal lattice, instead of a varying direction as in the Casimir method. The second is the very great sensitivity of the method; thus Pippard could detect changes of penetration as small as $0\cdot02 \times 10^{-6}$ cm., and was consequently able to follow the variation of penetration depth down to much lower temperatures than Laurmann and Shoenberg, whose limit of detection was only $0\cdot2 \times 10^{-6}$ cm. Against these advantages must be set the serious difficulty of interpretation. Except in the region of the lowest temperatures λ' differs from λ, and the theory accounting for the difference in detail is both complicated and of uncertain validity (since, as we shall see in § 6.3.3, special assumptions are involved). Ideally, what is wanted is a method of measurement at some intermediate frequency, high enough to permit of good accuracy, yet not so high as to produce any large difference between λ' and λ.

5.3.4. *High-frequency resistance method.* It is convenient to mention here that in principle measurements of the resistance of a superconductor at high frequencies (such as those of Pippard (1947a, 1950a) described in §1.4.6) also provide a method of deducing λ for macroscopic specimens. We shall see in §6.3.3 that the resistance depends on λ, but we shall see also that the theory involves special assumptions of uncertain validity, and predicts certain features which are not in agreement with experiment. Thus until the theoretical situation is improved this can hardly be regarded as a reliable method of finding λ to better than an order of magnitude.

5.4. Dependence of λ on crystal orientation

Ginsburg (1944) and von Laue (1947, 1948) pointed out that in general it was to be expected that λ would depend on the direction of current flow relative to the axes of the crystal lattice. In the preliminary experiments of Laurmann and Shoenberg (1947) the large differences between λ for different mercury specimens suggested that there was such an effect and that the crystal orientation varied from one specimen to another. Their later and more systematic experiments (1949) showed, however, that the originally observed large differences were more probably associated with rough surface conditions than with differences of crystal orientation. As already stated in §5.3.1, no appreciable difference could be found for tin between λ_1 and λ_3, the penetration depths for current flow perpendicular and parallel to the principal axis respectively, and only a slight difference was found for mercury. Since the current flows round the circumference of the cylinder, for a cylinder in which the principal crystallographic axis is perpendicular to the cylinder axis, the angle θ between the current and the principal axis varies round the circumference. Fraser and Shoenberg (1949) on the basis of a generalization of the Londons' theory (see §6.2.4) have shown that this should lead to observation of an 'effective' value of λ which is very nearly $\frac{1}{2}(\lambda_1 + \lambda_3)$, while for a cylinder with its axis parallel to the principal crystallographic axis it is evident that $\lambda = \lambda_1$. Thus, in the Casimir method, the maximum difference of λ available for observation is only half that between λ_1 and λ_3. It will be seen that the slight difference found for mercury, although corresponding to about 20% between λ_1 and λ_3, was really only a difference of about 10% in the measured values, and this is so little more than the uncertainties due to possible variations of surface conditions and measuring errors, that it is difficult to be certain that the effect is real.

Pippard (1950a), using the method outlined in §5.3.3 in which, as already mentioned, θ has a unique value for each specimen, has

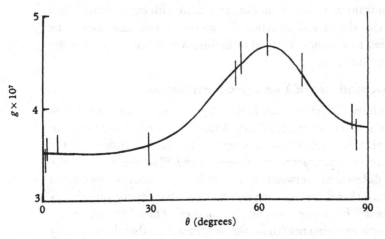

Fig. 66. Variation of penetration depth with crystal orientation for tin single crystals (Pippard, 1950a); g is a scaling factor proportional to λ_0, θ is the angle between the current and the tetragonal axis.

studied the variation of λ_0 with θ for tin single crystals (at 9200 mcyc./sec.). His results are shown in fig. 66, and give definite evidence of a variation, but one that obviously contradicts the relation

$$\lambda = \lambda_1 + (\lambda_3 - \lambda_1)\cos^2\theta \qquad (5.15)$$

which is to be expected on the basis of the Londons' theory (see §6.2.5). The fact of the variation itself does not conflict with the negative result found by Laurmann and Shoenberg, since in their experiments only two orientations were studied for tin, and the expected difference would have been between Pippard's result for $\theta = 90°$ (which should correspond to Laurmann and Shoenberg's result for a specimen with its tetragonal axis parallel to the cylinder axis) and some sort of average over θ of Pippard's results (for the perpendicular specimen of Laurmann and Shoenberg). It can be seen that this difference might well be too small to have been observed by the Casimir method. The disturbing aspect of the disagreement with (5.15) is that this relation is a rather direct consequence of the Londons' theory, so that if (5.15) is disproved, suspicion is cast on the validity of the Londons' theory, which, as we shall see in Chapter VI, in other respects seems to describe the facts well. It is not quite out of the question that the observed variation with θ is due to some subsidiary cause which has not yet come to light. It should be noticed that if (5.15) has to be replaced by some more general relation, the direction of the binary axis as well as that of the tetragonal axis might be involved. Now in Pippard's experiments the direction of the binary axis was not controlled, so it is perhaps a puzzling circumstance that a smooth curve against θ should be obtained. The fact also that the theory finds some support in Laurmann and Shoenberg's measurements on mercury makes the discrepancy for tin all the more surprising. More experiments on this rather crucial question are evidently needed.

5.5. Field dependence of λ

Until now we have tacitly assumed that λ is a function of temperature only, but as Ginsburg (1947) first pointed out, it is quite possible that it might depend also on field. In this section we shall discuss briefly the experimental evidence on this point, bearing in mind the possibility that the nature of such a field-dependence may be quite different according as the specimen is large or small compared with λ. The theoretical background of the problem will be discussed in §6.3.5.

For large specimens the best evidence is that of Pippard (1950c), who studied the changes in λ' as a transverse field was applied to a tin single-crystal wire about 10^{-2} cm. in radius. Owing to the effects discussed in Chapter IV, superconductivity is not destroyed until the maximum local field at the surface of the wire appreciably exceeds H_c, and this local field is, moreover, tangential and therefore in the same line as the field of the radio-frequency currents in the specimen. Pippard found that λ' changed by only about 2 % when the maximum local field was increased from zero to H_c, thus demonstrating that in a macroscopic specimen the penetration depth does not vary appreciably with field. A similar lack of variation of λ, but only up to $0.8H_c$, was shown also in the experiments of

Laurmann and Shoenberg (1949), but for technical reasons it was not possible to make measurements closer to H_c.

For specimens of size comparable with λ there is experimental evidence from the magnetization curves of Shoenberg (1940a) and of Lock (1951), but the interpretation is uncertain. Thus the fact that the magnetization curves of fig. 58 depart from linearity as the field is increased might be taken to mean that λ was increasing, but could equally well be attributed to the range of sizes of spheres present in the specimen. For Lock's thin films the evidence is slightly more convincing; fig. 67 shows magnetization curves of films of various thicknesses, and it can be seen that the 'rounding-off' sets in earlier and is generally more pronounced the thinner the specimen or the higher the temperature (i.e. the larger is λ/a). Although the film thicknesses were much more definitely determined than was the sphere radius in the colloid experiments, it is by no means certain that there were not variations in thickness which were relatively more pronounced for the thinner films. The rounding at higher temperatures may also be partly due to the spread of critical field associated with a spread of the transition temperature. The shape and area of the magnetization curves of colloids and thin films will be further discussed in §§ 5.8 and 6.3.5.

We may mention here that a particular kind of field dependence of λ is predicted by Heisenberg's theory of superconductivity (see § 6.4). In this theory it is supposed that a superconductor can in principle carry only a certain maximum current density which falls off rapidly as the temperature approaches absolute zero; this leads to the result (Koppe, 1950) that for a thin film λ should increase more and more with field as the temperature is lowered, since only in this way can the current density at a given field be kept as low as necessary. This means in turn that the rounding of the magnetization curve should get more and more marked as the temperature is lowered, particularly for thin films and for high values of T_c. In some special measurements with thin lead films Lock (1951) was unable to find any evidence for such an effect.

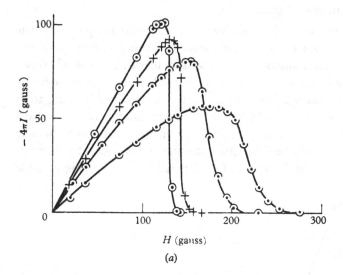

H (gauss)

(a)

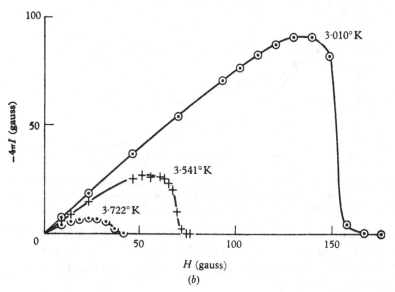

H (gauss)

(b)

Fig. 67. Magnetization curves of thin tin films (Lock, 1951). (a) Various thicknesses at $3°$ K.: $\odot$ 79×10^{-6} cm., $+$ 55×10^{-6} cm., $\odot$ 30×10^{-6} cm., $\odot$ $23 \cdot 2 \times 10^{-6}$ cm. (b) Various temperatures for a thickness of $37 \cdot 6 \times 10^{-6}$ cm.

5.6. The law of penetration

As explained in § 5.2, evidence for or against any particular law of penetration can come only from experiments with specimens whose size is comparable with or smaller than λ. The colloid experiments of § 5.2.1 are of little use, since the absolute value of the radius a was not known, but some evidence was obtained by comparing results for a number of different colloids of different (but unknown) average particle size. For each preparation the curve of κ/κ_0 against T could be translated into one of κ/κ_0 against a quantity proportional to λ. It was found possible, by choosing suitable

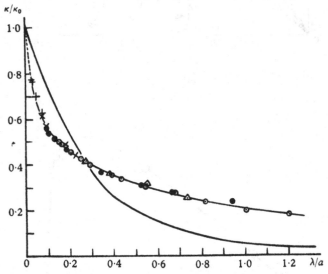

Fig. 68. Variation of κ/κ_0 with λ/a deduced from a variety of colloid preparations (Shoenberg, 1940a). The scale of the abscissae for the experimental points has been arbitrarily chosen; the full curve is based on equation (5.16).

scaling factors for the abscissae, to fit all the curves into a single curve which was interpreted as a curve of κ/κ_0 against λ/a, with some arbitrary scale of abscissae. Now the form of this curve is predicted by the Londons' theory as

$$\frac{\kappa}{\kappa_0} = 1 - \frac{3\lambda}{a}\coth\frac{a}{\lambda} + \frac{3\lambda^2}{a^2} \qquad (5.16)$$

(see Appendix II), which, as can be seen from fig. 68, disagrees badly with the experimental results. In view of the wide range of particle sizes present in each of the colloids this disagreement may be due simply to the distortion of (5.16) produced by averaging over the size distribution, but in the absence of any knowledge of the size distribution, this interpretation (originally thought improbable) must be regarded as speculative, and the possibility of a real discrepancy must be borne in mind (see, however, Pippard, 1952).

For the plates of Lock's experiments the prediction of the Londons' theory (see Appendix II) is

$$\frac{\kappa}{\kappa_0} = 1 - \frac{\lambda}{a}\tanh\frac{a}{\lambda}. \tag{5.17}$$

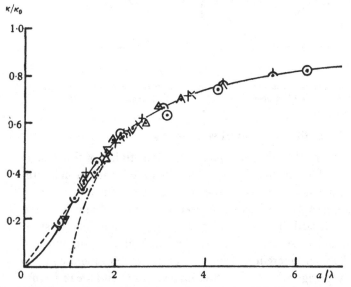

Fig. 69. Variation of κ/κ_0 with a/λ (assuming $\lambda = \lambda_0(1-(T/T_c)^4)^{-\frac{1}{2}}$ with $\lambda_0 = 5\cdot0 \times 10^{-6}$ cm.) for various thin tin films (Lock, 1951). The chain curve is based on equation (5.5) and the broken curve on the penetration law illustrated in fig. 70 ($\kappa/\kappa_0 = 1 - \lambda/a$ for $a \geqslant 2\lambda$, $\kappa/\kappa_0 = a/4\lambda$ for $a \leqslant 2\lambda$).

It can be seen from fig. 69 that if the values of κ/κ_0 are plotted against $a\{1-(T/T_c)^4\}^{\frac{1}{2}}/\lambda_0$ with $\lambda_0 = 5\cdot0 \times 10^{-6}$ cm., the points for all the specimens lie well on a single curve, which has the form (5.17). This seems to provide confirmation of the truth of (5.17), but the agreement is not so impressive as it appears at first sight. In the first place, as can be seen from the chain line, it is only for the thinnest films at the highest temperatures that (5.17) differs appreciably from its limiting form (5.5), valid for thick films. Secondly, Lock has shown that the form of the theoretical curve is not very sensitive in the relevant range to the precise penetration law assumed; thus on the assumption of the penetration law illustrated in fig. 70, the predicted variation of κ/κ_0 would follow the broken curve, which fits the points almost as well as does (5.17). Thus the strongest conclusion that can be drawn from Lock's data is that they are not inconsistent with the Londons' penetration law.

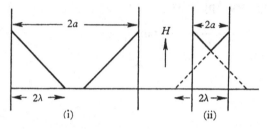

Fig. 70. Illustrating the hypothetical penetration law from which the broken curve of fig. 69 has been derived (Lock, 1951), (i) $a \geqslant 2\lambda$. (ii) $a \leqslant 2\lambda$.

5.7. Critical fields of small specimens

5.7.1. *Critical fields of thin films.* The first studies of thin superconducting films to give significant results were resistance measurements, carried out at about the same time by Shalnikov (1938, 1940a) and Appleyard and co-workers (Appleyard and Misener, 1938; Appleyard, Bristow and London, 1939; Appleyard, Bristow, London and Misener, 1939).* In this work films of tin and lead (Shalnikov) and mercury (Appleyard *et al.*) were evaporated on to glass in very rigorous conditions of high vacuum, and the resistance studied as the temperature was reduced. It was found that full superconductivity set in sharply at about the usual transition temperature of the bulk metal if the film had been suitably annealed (though without annealing the transition temperature was sometimes as much as one degree higher†), even for films as thin as

* The earlier work of Burton, Wilhelm and Misener (1934), Misener and Wilhelm (1935), Misener, Smith and Wilhelm (1935) and Misener (1936) on electrolytically deposited films will not be described, since many of the complicated features of their results were probably due to alloying rather than size effects. A general account of this earlier work may be found in the first edition of this monograph (Chapter VII).

† The behaviour of unannealed tin films has been further studied by Buckel and Hilsch (1952).

5×10^{-7} cm. (i.e. only 15 atomic layers thick). The most important result of these experiments was that the critical fields (h) of the thin films were very much higher than the bulk critical fields (H_c), and for mercury a detailed study of the variation of h/H_c with size and temperature was made. The results are illustrated in fig. 71 a. For thick films a relation of the form

$$h/H_c = 1 + b/2a, \qquad (5.18)$$

was well obeyed at any one temperature; the constant b was about 12×10^{-6} cm. at low temperatures, and showed a temperature variation similar to that of λ.

Alekseyevsky (1941 b) has studied the destruction of super-conductivity in thin tin films ($1\cdot5 \times 10^{-6}$ to $2\cdot4 \times 10^{-5}$ cm.) evaporated in good vacuum conditions on to glass; he used the ingenious method of suspending the film from a fine quartz fibre and observing the field strength at which the couple holding the plane of the film nearly in the field direction disappeared.* The temperature variation of h/H_c for the various films found in this way is shown in fig. 71 b, and can be seen to have a character similar to that found by Appleyard et al. for mercury (fig. 71 a), though Alekseyevsky's thickest films were too thin for (5.18) to apply. Unfortunately, for technical reasons, he could not obtain data much above $3\cdot0°$ K., so the rise of h/H_c as T_c is approached is not as evident as with the mercury films.

The interpretation of the critical field data shown in fig. 71 for the thinner films, and the meaning of the parameter b, will be discussed after the necessary theory has been developed in § 5.8.

---

* The interpretation of the results is complicated by the presence of a small normal component of the field, which gets smaller as the applied field is increased; owing to the hysteresis features in the magnetization curve of a thin plate in a transverse field (see fig. 46, p. 119), the couple may seem to vanish at a field smaller than that which destroys superconductivity if the plate is initially subjected to too large a normal component of field. Alekseyevsky mentions however, that his critical field values are in fair agreement with those of Shalnikov, and also for a specially prepared thick film (rolled) they agree with bulk values. This suggests that the complication mentioned was not in fact relevant, though it may have contributed to the scatter of the experimental points; the details of the experimental procedure given in the published paper are not sufficient to decide this question. Since no data were given to indicate the values of T_c for the individual films, it is possible that systematic errors may be present in the points plotted in fig. 71 b, which have been calculated from Alekseyevsky's data on the assumption that T_c was the same as for bulk tin.

Using relatively thick rolled films of tin, Andrew (1949) also verified the relation (5.18), and found a temperature variation of b of the same general character as that just described; the value of b at $3 \cdot 0°$ K. was about 15×10^{-6} cm. Lock (1951), too, found high critical fields from his magnetic measurements, and as can be seen

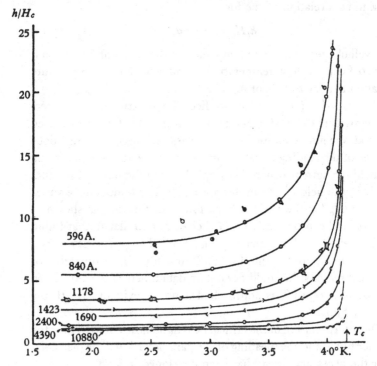

Fig. 71a. Temperature variation of h/H_c for mercury films of various thicknesses (Appleyard, Bristow, London and Misener, 1939); the thicknesses are given in Angstroms.

from fig. 72, (5.18) is again approximately satisfied for the thickest films; he found values of b of 13×10^{-6} cm. for tin at $3 \cdot 0°$ K., 11×10^{-6} cm. for indium at $2 \cdot 0°$ K. and 9×10^{-6} cm. for lead at $4 \cdot 2°$ K. In view of the shifts of T_c and the smearing of the transitions, the values of H_c used in reducing the data were rather uncertain, so the estimate of b for tin can be considered as in fair agreement with that of Andrew; the estimate for lead is in fair agreement with an estimate based on data by Pontius for thin wires (see § 5.7.2).

5.7.2. *Critical fields of thin wires.* The first indication of an increase of critical field with diminishing size came from resistance measurements of Pontius (1937) on thin lead wires (down to radius 5×10^{-4} cm.). He, too, found a variation of the form (5.18); here, however, it is convenient to modify the form of (5.18) to

$$h/H_c = 1 + b/a, \qquad (5.19)$$

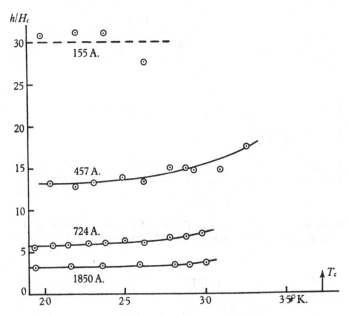

Fig. 71*b*. Temperature variation of h/H_c for tin films of various thicknesses (Alekseyevsky, 1941*b*). To avoid confusion the data for the thickest film have been omitted; for technical reasons no data could be obtained close to T_c; the thicknesses are given in Angstroms.

where a is the radius of the wire, in order that b shall have the same interpretation for the wire as it did in (5.18) for a plate (cf. (5.5) and (5.6), and (5.26) and (5.27)). If Pontius's data are fitted to (5.19) a value of $b = 11 \times 10^{-6}$ cm. is found at $4.2°$ K., which, bearing in mind that the largest effect found was an increase of only 4%, agrees well with the value found by Lock.

5.7.3. *Critical fields of colloids.* The magnetization curves of the colloids discussed in §5.2.1 are shown continued to much higher fields in fig. 73. It will be seen that the critical fields are indeed very

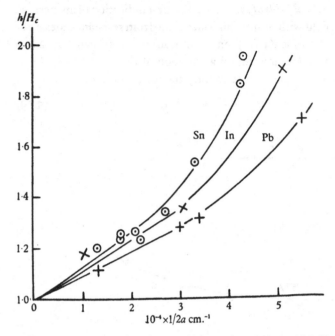

Fig. 72. Variation of h/H_c with reciprocal thickness for relatively thick films of indium (at 2° K.), tin (at 3° K.) and lead (at 4·2° K.) (based on Lock, 1951).

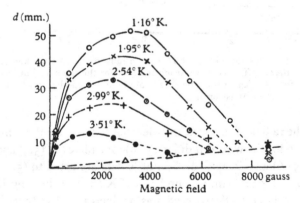

Fig. 73. High field magnetization curves of the mercury colloid the behaviour of which at lower fields is shown in fig. 58 (Shoenberg, 1940a). As in fig. 58 the galvanometer deflection d is proportional to magnetization; for technical reasons the chain line, rather than the axis of abscissae, is the true base line.

high, and the temperature variation of h/H_c is shown in fig. 74. It must not be forgotten that while κ/κ_0 referred to the *average* particle size, h/H_c always refers to the *smallest* particle size which is present in sufficient strength to have an appreciable magnetization. Here, again, it is convenient to defer discussion of the results until the theory has been developed (§§ 5·8 and 6.3.5).

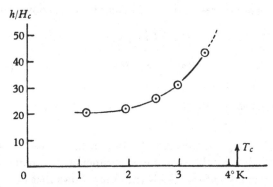

Fig. 74. Temperature variation of h/H_c for the colloid whose magnetization curves are shown in figs. 58 and 73 (Shoenberg, 1940a).

5.8. Thermodynamic considerations*

In this section we shall show how thermodynamics provides a qualitative explanation of the increased critical fields of small specimens, and shall indicate some of the difficulties of making the theory quantitative. As was pointed out in § 3.1, for a body of any shape or size we have

$$G_n - G_s = -V \int_0^{H_e} I \, dH_e = V\mathscr{A}, \qquad (5.20)$$

where $\mathscr{A}$ is the area under the magnetization curve and H_e is a field big enough to destroy superconductivity. Since the slope of the magnetization curve is less for a small specimen than for a bulk specimen of the same shape, it is clear that the magnetization curve will have to continue to higher fields to have the same area, i.e. the critical field will be higher. This qualitative argument assumes, however, that $(G_n - G_s)/V$ is independent of size, i.e. equal to

* The discussion of this section is based essentially on work by Pippard (1951a), and I am indebted to him for informing me of it in advance of publication.

$H_c^2/8\pi$, and this cannot be taken for granted. To take account of possible modifications of the free energy close to the surface, we can introduce surface free energy contributions (cf. p. 126) and write

$$G_n - G_s = VH_c^2/8\pi + A(\alpha_n - \alpha_s),\qquad(5.21)$$

where A is the surface area of the specimen, and α_n and α_s the surface-free energies per unit area of the free surface in the normal and superconducting phases respectively. This can be more conveniently written as

$$\frac{G_n - G_s}{V} = \frac{H_c^2}{8\pi}\left(1 + \frac{\beta}{d}\right),\qquad(5.22)$$

where

$$\beta = 8\pi(\alpha_n - \alpha_s)/H_c^2,\qquad(5.23)$$

and d is V/A, i.e. the characteristic small dimension of the specimen. Equation (5.20) then becomes

$$\mathscr{A}/\mathscr{A}_0 = 1 + \beta/d,\qquad(5.24)$$

if we write $\mathscr{A}_0$ for $H_c^2/8\pi$, the area of a bulk magnetization curve. Comparison of this result with the experimental magnetization curves of figs. 73 and 67 for colloids and films shows that in fact β is very small and may well be zero. The tin films suggest a value of about 4×10^{-6} cm., but the argument has assumed reversible processes, although actually there is considerable hysteresis, so this value is really an upper limit. The mercury colloid results, which suggest values of β ranging from -3×10^{-7} to 1×10^{-7} cm. are perhaps more significant, since there was no hysteresis in the magnetization curves and, moreover, the size (i.e. d) was much smaller.

From (5.24) the critical field of a film or a wire parallel to the applied field can be deduced if further assumptions are made about the magnetization curve. In particular, if it is assumed that the magnetization curve is linear with slope κ right up to a critical field h, then $\mathscr{A} = \frac{1}{2}\kappa h^2$, so that

$$\kappa h^2/\kappa_0 H_c^2 = 1 + \beta/d,$$

or

$$(h/H_c)^2 = (1 + \beta/d)/(\kappa/\kappa_0).\qquad(5.25)^*$$

For the particular case of fairly thick specimens where (5.5) or (5.6) applies, this reduces to

$$h/H_c = 1 + (\lambda + \beta)/2a \quad\text{for a film of thickness } 2a,\qquad(5.26)$$

and

$$h/H_c = 1 + (\lambda + \beta)/a \quad\text{for a wire of radius } a.\qquad(5.27)$$

* A relation equivalent to this was first obtained by Ginsburg (1945).

Now we have seen in § 5.7 that for both films and wires a relation of the form (5.26) or (5.27) has been found experimentally. If the empirical parameter b is identified with $\lambda + \beta$, the value of β deduced for tin is about 8×10^{-6} cm. at low temperatures and varies with temperature in much the same way as does λ. Such a high value of β is, however, quite inadmissible in view of the evidence from the areas of the magnetization curves, and so the argument leading to (5.26) and (5.27) must be unsound. Again, if β is *assumed* to be negligible, and h/H_c is calculated for thinner films by means of (5.25), using values of κ/κ_0 based on (5.17), and using the best available information about λ for mercury, we find values a good deal smaller than the experimental values of Appleyard *et al.* (see fig. 71).

The fallacy lies in assuming that the magnetization curve has a constant slope all the way to the critical field which restores resistance, and, indeed, as can be seen from fig. 67, the experimental magnetization curves of Lock's films are appreciably rounded. For the thinner films part of this rounding may (as mentioned in § 5.6) be due to an increase of penetration depth with field (see § 6.3.5 for a theoretical discussion), but this interpretation is not very plausible for the thicker films in view of Pippard's result that λ is practically independent of field. Another possibility is that the falling off of magnetization is due to appearance of normal 'islands' below the critical field, and as we saw in § 4.6.2, this can happen only if Δ, the parameter associated with the interphase surface energy, is negative. This may happen at places of inhomogeneous strain (as in Faber's superconducting nuclei (see § 4.6.3)), but Pippard has suggested that it may also happen in thin films for more fundamental reasons (see § 6.3.5).

The situation may be summarized, then, by saying that the critical fields of thin films are higher than $(\kappa_0/\kappa)^{\frac{1}{2}}$ times H_c (i.e. higher than indicated by (5.25) with $\beta = 0$), simply because, for various possible reasons, the slope of the magnetization curve falls off from its initial value κ as the field is raised. The precise nature of this falling off has not yet been quantitatively discussed, though qualitative discussion by Pippard suggests that the empirical relation (5.18) for fairly thick films is in reasonable accord with the hypothesis of normal island formation below the critical field. The detailed features of the results of Appleyard *et al.* on thinner films have not yet found any complete interpretation.

For small spheres it can be easily seen that if β is neglected, and if it is assumed again that the magnetization curve is linear, with slope κ right up to the field h which destroys superconductivity, then the appropriate modification of (5.25) is

$$h/H_c = (\tfrac{2}{3}/(\kappa/\kappa_0))^{\frac{1}{2}}. \qquad (5.25\,a)$$

For the colloid whose magnetization curves are shown in figs. 58 and 73, this predicts values of h/H_c of about 9 at low temperatures and 12 at $3\cdot5°$ K., which are much lower and show less relative variation with temperature than the experimental values 22 and 42 respectively. Since, however, the experimental values of h/H_c refer to the *smallest* spheres in the colloid while the κ/κ_0 values are averages over all sizes, this discrepancy does not necessarily disprove the applicability of (5.25 a). A modification of the theory taking into account a possible variation of penetration depth with field for very small particles will be outlined in §6.3.5, and we shall see that this modification offers some possibility of improving the agreement between theory and experiment as regards the temperature variation of h/H_c.

5.9. Critical currents in thin specimens

There is very little experimental evidence on how the criterion for destruction of superconductivity by a current (i.e. the Silsbee hypothesis) is modified if the dimensions of the wire carrying the current are reduced. It will be convenient to discuss the theoretical aspects of this problem before surveying such experimental evidence as there is.

The method of approach of §5.8 cannot be directly applied except perhaps in the limiting case of wires so thin that the current is uniformly distributed across the cross-section (see below).* Thus, equating the appropriate free energies of the wire in the super-

---

* The application of thermodynamics to this problem is of uncertain validity because the current causes continuous production of entropy in the normal state. Since, however, it is possible to obtain correct results in other problems (e.g. thermoelectric effects) involving non-equilibrium states, by ignoring such entropy production, it is not unlikely that here too we may ignore the Joule heat and regard the state of the normal metal carrying a current as one of thermodynamic equilibrium.

conducting state carrying just the critical current, and of the wire in the normal state carrying the same current, is not permissible for a macroscopic wire because the transition is irreversible; energy is dissipated as the current and its magnetic field spread into the interior of the cross-section, and this process cannot be carried out infinitely slowly since it is intrinsically unstable (see Pippard, 1950d). In any case, as we saw in §4.7, there is no unique critical current at which the whole wire goes into the normal state, but merely a critical current at which resistance begins to reappear. This critical current can be expressed in a way which is rather more fundamental than the Silsbee hypothesis, as was first pointed out by H. London (1935). He considers a layer much thinner than λ, containing a boundary between superconducting and normal material and discusses the free-energy balance during a reversible isothermal displacement of this boundary. Since the displacement is reversible the increase of Helmholtz free energy of the layer is equal to the work done on the layer by the sources of the magnetic field irrespective of any heat flow resulting from the magneto-caloric effect. The magnetic field and the electric field induced by the movement of the boundary are both tangential to the boundary, and since the layer is very thin have the same values on either side of the layer; thus the Poynting vector has the same value on both sides of the layer so that the work done by the sources of the field is zero. Consequently, the movement of the boundary causes no change of free energy. If f_s and f_n are the local values of the Helmholtz free energies per unit volume of the superconducting and normal phases respectively in zero field and J is the current density at the superconducting boundary, this must mean that

$$f_n - f_s - \frac{1}{2}\left(4\pi\frac{\lambda^2}{c^2}\right)J^2 = 0, \qquad (5.28)$$

since, as we shall see in §6.2.3, $\frac{1}{2}(4\pi\lambda^2/c^2)J^2$ is the extra energy per unit volume of the superconducting state due to the kinetic energy of the 'superconducting' electrons (over and above the field energy terms $(H^2 + E^2)/8\pi$, which have the same values immediately on either side of the boundary).

Now
$$f_n - f_s = H_c^2/8\pi,$$

so we find
$$J = cH_c/4\pi\lambda. \qquad (5.29)$$

According to this argument, then, the criterion for destruction of superconductivity is that there is a critical *current density* given by (5.29), rather than a critical *field*.

For a wire of radius a, where $a \gg \lambda$, the highest current density is $i/2\pi a\lambda$, so (5.29) gives

$$i = \tfrac{1}{2}acH_c, \tag{5.30}$$

i.e. the same answer as Silsbee's hypothesis. There is, however, nothing in London's argument which limits the application of (5.29) to thick wires, and we might expect that for very thin wires ($a \ll \lambda$), where J is uniform, and given by $i/\pi a^2$, we should have

$$i = a^2 cH_c/4\lambda. \tag{5.31}$$

Similarly for a very thin superconducting film of thickness d, where $d \ll \lambda$, coated on a cylinder of radius a, where $a \gg \lambda$, $J = i/2\pi ad$, so

$$i = adcH_c/2\lambda. \tag{5.32}$$

Both (5.31) and (5.32) are, of course, smaller currents than would be given by Silsbee's hypothesis.

It is, however, not certain that London's argument can really be applied to very thin wires on account of the 'long-range' order within a superconductor (see §6.3.5).* Thus Pippard (1951a) has suggested that in specimens of a size small compared with the range of this order ($\sim 10^{-4}$ cm.) it is impossible to conceive of a boundary between normal and superconducting phases, so that the body has to be in either the one or the other state. If we consider a very thin wire carrying a current, the field distribution is practically the same whether it is normal or superconducting, so the change from one state to the other becomes practically reversible, and the objection to equating free energies no longer applies. We then have

$$f_s + \frac{1}{2}\left(4\pi \frac{\lambda^2}{c^2}\right)J^2 = f_n + \tfrac{1}{2}\Lambda_n J^2. \tag{5.33}$$

Here we have ignored surface energies in accordance with the indications of §5.8; the term in $H^2/8\pi$ comes equally on both sides of the equations (since the wire is assumed so thin that the current distribution is uniform in both normal and superconducting states) and so has been omitted. The extra term $\tfrac{1}{2}\Lambda_n J^2$ represents kinetic energy of the normal electrons;† for a macroscopic wire this provides a negligible contribution compared with the super-

---

* I am indebted to Dr A. B. Pippard for the following argument.

† The value of Λ_n can easily be shown to be τ/σ, where τ is the relaxation time of the electrons and σ the conductivity in the normal state of the metal.

conducting term in J^2 (when integrated over the cross-section of the wire), but for a very thin wire they are quite comparable. Thus

$$J = cH_c/4\pi\lambda', \qquad (5.34)$$

where $\qquad \lambda' = (\lambda^2 - c^2\Lambda_n/4\pi)^{\frac{1}{2}}. \qquad (5.35)$

This argument suggests, then, that the critical currents for thin wires or thin films may not be as small as indicated by (5.31) or (5.32). It is, however, not clear how far it is permissible to apply thermodynamic arguments when in one of the states considered (here, the normal state carrying a current) energy dissipation is an intrinsic feature.

So far there has been no experimental work on thin wires owing to the difficulties of preparing thin enough wires. Attempts have been made to study the critical current in thin films deposited on macroscopic cylinders; in the work of Misener (1936) critical currents much lower than indicated by the Silsbee hypothesis were indeed found in electrolytically deposited films, but in view of other features of the results it is far from certain that this was purely a size effect (for instance, Silsbee's hypothesis was still not observed for films as thick as 3×10^{-4} cm., i.e. much thicker than λ). More recently Lock (unpublished) has measured critical currents of tin films deposited on quartz by evaporation and also found low values; from other indications it seemed probable, however, that these low values were more due to a broken-up structure of the film than to a genuine size effect. The only experiments in which the films may have been satisfactory as regards structure were those of Shalnikov (1938, 1940a) (see also § 5.7.2), but unfortunately his films were deposited on plane surfaces, so that the interpretation is much less certain. The difficulty is that in a thin strip the current is concentrated in the edges, and the difficult mathematical problem of finding the current distribution according to the Londons' equation has not yet been solved for the general case of a thin strip (either of rectangular or elliptical cross-section). In two limiting cases, however, an approximate solution can be given. If the strip is treated as having an elliptic cross-section of minor axis $2a$ and major axis $2b$ and $a \gg \lambda$, then, as we saw in § 2.9.2, the highest current density is at the end of the major axis, and (2.27) shows immediately that the maximum current should be

$$i = \tfrac{1}{2}acH_c. \qquad (5.36)$$

If, on the other hand, the strip is sufficiently thin,* the current density must be uniform and given by

$$J = i/\pi ab.$$

If we can assume that (5.29) still applies, we find the maximum current should be

$$i = abcH_c/4\lambda. \tag{5.37}$$

The only relevant data given by Shalnikov are for two annealed tin strips and two annealed lead strips, and a comparison of Shalnikov's results with (5.36) and (5.37) is shown in Table VI.† It is at least satisfactory that the observed values lie between the limiting predictions of (5.36) and (5.37), and it is perhaps significant that the ratio of the observed value to the prediction of (5.36) is least for the largest value of a/λ. One might expect that this ratio would increase as the temperature rises, since a/λ is then smaller, but there is no reliable data sufficiently close to T_c to produce a significant change of λ.

TABLE VI

(See Shalnikov, 1940a)

Metal	$a \times 10^6$ cm.	$b \times 10^2$ cm.	Temp. (° K.)	a/λ	i (mA.) from (5.36)	i (mA.) from (5.37)	i (mA.) observed
Sn	9·2	2	2·0	2·0	9·7	1930	120
Sn	2·7	2	2·0	0·54	2·8	560	50
Pb	9·4	2	2·0	2·4	35	8700	300
Pb	2·6	2	4·2	0·65	7·3	1800	110

Notes. No precise values of b are given for the individual films, but the description of the method mentions that the width, $2b$, was usually about 4×10^{-2} cm. In the calculations we have assumed H_c to be 210 gauss for Sn and 750 gauss for Pb at 2° K. and 560 gauss for Pb at 4·2° K. In estimating a/λ we have taken λ as 5×10^{-6} cm. for Sn at 2·0° K. and 4×10^{-6} cm. for Pb at both 2° K. and 4·2° K. (see Table V).

* It is probable that 'sufficiently thin' means $ab < \lambda^2$, which would give an impossibly small value of a.

† It may be mentioned that before annealing, the critical currents were much smaller and of the same order as predicted by (5.36); this, however, is probably a coincidence, since (as mentioned in § 5.7.1) the transition temperatures were appreciably higher, suggesting that considerable stresses were present and that the films could not be regarded as differing from bulk material merely as regards thickness. In some respects the behaviour of the unannealed films was similar to that of alloys, as was the case for electrolytically deposited films (see footnote, p. 166).

CHAPTER VI

THEORETICAL ASPECTS

6.1. Introduction

The theory of superconductivity can be considered in several stages. The first stage is to accept the macroscopic description of a superconductor as a body with $B = 0$ and $E = 0$ and to develop the electrodynamic and thermodynamic consequences as far as is possible without making further assumptions. This has already been done in Chapters II and III, where it has been shown that quite a variety of observed effects can be correlated. Introduction of surface energies permits some further advances in dealing with problems of mixtures of the normal and superconducting phases, as in the intermediate state, and questions of the kinetics of the transition. Here already the theory is on less certain ground, since assumptions have to be made about the behaviour of very small regions, and moreover, as we have seen in Chapter IV, mathematical difficulties prevent a rigorous working out of the theory even on the basis of the assumptions made.

In this chapter we shall be concerned mostly with two further theoretical advances—that of F. and H. London (1935 a, b) in formulating a more general description of the electrodynamic behaviour and that of Gorter and Casimir (1934 b), who introduced a kind of two-fluid model, rather like that which has been so successful in the theory of liquid helium, in order to describe the thermodynamic behaviour in a slightly more fundamental way. Both these theories are still of the 'phenomenological' kind, since in each of them an arbitrary postulate is made in order to fit the facts, but they are valuable not only in suggesting how to interpret the experimental data but also in suggesting further experiments. It should be emphasized, however, that neither theory can be regarded as experimentally proved, and that the Gorter-Casimir picture is indeed difficult to reconcile with recent results at very high frequencies (see §6.3.4), so that they must not be regarded as infallible guides.

The last stage of the theory—which is still far from complete, though there has been considerable progress recently—is to find the fundamental reasons why the postulates of the phenomenological theories work as well as they do. In this stage the theory has to go beyond the conventional electron theory of metals and show just what interaction between the electrons produces the superconducting transition, and why the metal has superconducting properties once the transition has occurred. Here the phenomenological theories, particularly the Londons' theory, are again valuable in simplifying the formulation of the problem.

A full discussion of the Londons' theory can be found in several detailed texts (F. London, 1950; von Laue, 1949; Ginsburg, 1946), and in this chapter we shall deal with it only briefly (§ 6.2), emphasizing the basic ideas and quoting only such consequences of it as are relevant to the experimental results described in the previous chapters. This will be followed (§ 6.3) by an account of the Gorter-Casimir theory showing how it enables the temperature variation of penetration depth to be correctly predicted, and pointing out both its successes in providing qualitative description of various effects and its failures in certain quantitative respects. After an account of some evidence, due to Pippard, for long-range order in the superconducting state (§ 6.3.5), the chapter concludes (§ 6.4) with a brief qualitative account of recent attempts at a fundamental theory. We shall see that the theories of Heisenberg (1947, 1948 a, b) and of Born and Cheng (1948) have several serious failings, both intrinsic and as regards agreement with experiment, but that the recent theories of Fröhlich (1950) and Bardeen (1950) are more promising.

6.2. The Londons' theory*

6.2.1. *Formulation of the theory.* The Londons' theory (F. and H. London, 1935 a, b) grew out of an older theory of Becker, Heller and Sauter (1933) which considered the superconductor as the limiting case of a perfect conductor. Becker, Heller and Sauter pointed out that if the electrons encountered no resistance, an applied electric field would accelerate them steadily. Thus if there

---

* It should be noted that throughout this section we use the description in which $B = H$, and the currents are explicitly recognized, rather than taking account of the magnetic properties by distinguishing between B and H (see § 2.4).

are n electrons per c.c. of mass m, charge e and velocity $\mathbf{v}$, we should have
$$m\dot{\mathbf{v}} = e\mathbf{E},$$

and since the current density $\mathbf{J}$ is given by
$$\mathbf{J} = ne\mathbf{v},$$

we have
$$\mathbf{E} = \frac{4\pi\lambda^2}{c^2}\dot{\mathbf{J}}, \tag{6.1}$$

where
$$\lambda^2 = mc^2/4\pi ne^2. \tag{6.2}$$

Taking curls on both sides of (6.1) and using Maxwell's equation $\mathrm{curl}\,\mathbf{E} = -\dot{\mathbf{H}}/c$, we find
$$\frac{4\pi\lambda^2}{c}\,\mathrm{curl}\,\dot{\mathbf{J}} + \dot{\mathbf{H}} = 0. \tag{6.3}$$

Substituting Maxwell's equation $\mathrm{curl}\,\mathbf{H} = 4\pi\mathbf{J}/c$, we obtain
$$\nabla^2\dot{\mathbf{H}} = \dot{\mathbf{H}}/\lambda^2. \tag{6.4}$$

The physical meaning of this differential equation is that the value of $\dot{\mathbf{H}}$ disappears exponentially inside the surface of a macroscopic piece of the perfect conductor.* The distance in which $\dot{\mathbf{H}}$ decays is of order λ, which, if one electron per atom is assumed in (6.2), is about 10^{-6} cm. In our previous discussions of the properties of a perfect conductor (§2.2) the electric field associated with the acceleration of the electrons was ignored, and so the simpler result $\dot{\mathbf{H}} = 0$ was obtained instead of (6.4). The novelty of (6.4) is in showing that the value $\dot{\mathbf{H}} = 0$ is to be found only at a depth inside the metal much greater than the penetration depth λ. If the electrons had no mass, λ would be zero, and (6.4) would reduce to $\dot{\mathbf{H}} = 0$. Just as $\dot{\mathbf{H}} = 0$ led on integration to $\mathbf{H} = \mathbf{H}_0$, where $\mathbf{H}_0$ could be arbitrary, in contradiction to the experimentally observed Meissner effect, so too (6.4) is contradicted by experiment. Integrating with respect to time, (6.4) becomes
$$\nabla^2(\mathbf{H} - \mathbf{H}_0) = (\mathbf{H} - \mathbf{H}_0)/\lambda^2, \tag{6.5}$$

where $\mathbf{H}_0$ is an arbitrary field (whatever field happened to be inside the body when it last lost its resistance). Since the Meissner effect specifies that the field well inside a superconductor is always zero, (6.5) cannot apply to a superconductor as it stands. The Londons suggested that since the macroscopic theory of a perfect conductor

* A general proof that this is a property of the differential equation (6.4) has been given by von Laue (1949, § 7*f*).

(i.e. ignoring the electron inertia as in Chapter II) makes correct predictions about superconductors for the special case $H_0 = 0$, it was reasonable to suppose that the Becker, Heller and Sauter theory, too, might be correct in its detailed predictions of the influence of electron inertia in determining the way the field fell off at the surface of a superconductor, provided H_0 was made zero. More specifically, they suggested the new equation

$$\frac{4\pi\lambda^2}{c}\operatorname{curl}\mathbf{J} + \mathbf{H} = 0, \tag{6.6}$$

which can be obtained by time integration from (6.3) only if it is assumed that the constant of integration is zero. Equation (6.6) leads immediately to

$$\nabla^2\mathbf{H} = \mathbf{H}/\lambda^2, \tag{6.7}$$

to which (6.5) reduces in the special case $H_0 = 0$, and also to

$$\nabla^2\mathbf{J} = \mathbf{J}/\lambda^2. \tag{6.8}$$

The assumption $H_0 = 0$ has, of course, been introduced arbitrarily to agree with experiment, but we shall see later that the exclusion of other possible values of H_0 is in accordance with the idea of the superconductor as a macroscopic 'quantum mechanism'.

Thus, the Londons' theory generalizes the equations

$$\mathbf{H} = 0, \quad \mathbf{E} = 0, \tag{6.9}$$

which as we saw in Chapters I and II provide an adequate description of the *interior* of a macroscopic superconductor, to

$$\mathbf{H} = -\frac{4\pi\lambda^2}{c}\operatorname{curl}\mathbf{J}, \tag{6.6}$$

$$\mathbf{E} = \frac{4\pi\lambda^2}{c^2}\dot{\mathbf{J}}, \tag{6.1}$$

and in this way determines the fields not only in the interior of a macroscopic specimen but also in the region close to the surface, and in specimens of dimensions comparable to λ. The question of the extent to which these two equations should be regarded as independent will be discussed in §6.2.2. As we have already seen, these two equations combined with Maxwell's equations lead to the differential equations (6.7) and (6.8) for the magnetic field and the current density. For large specimens, the characteristic feature of the solutions of these equations is that they decay exponentially into the interior of the specimen, so that field and current density are negligibly small except at distances comparable to λ from the surface of the specimen. Thus for macroscopic specimens, the Londons' theory provides a description of the Meissner effect and leads to predictions practically identical with those of Chapter II based on the cruder description $\mathbf{B} = 0$. For specimens of size comparable to λ or small compared with λ, there is not sufficient room for the field or current density to decay to zero, and the characteristic differences of magnetic behaviour appear which have been discussed in Chapter V. As has been explained in §5.6, the experimental evidence about the validity of (6.7) and (6.8) is still inconclusive, but even if they should prove not to be completely accurate, they provide a convenient and mathematically tractable description of the penetration and give results which are certainly qualitatively accurate. These equations have been solved for some simple special cases such as plates, cylinders and spheres in uniform magnetic fields, and cylinders carrying a current, and the results, some of which have already been quoted in Chapter V, are collected in Appendix II.

6.2.2. *The significance of the Londons' equations.* An important question is whether either of the London equations (6.1) or (6.6) is to be regarded as more fundamental than the other. These two equations describe the zero resistance and the Meissner effect respectively and it is evidently important to decide whether it is at all possible to say that one of these effects is a consequence of the other or whether they must be regarded as entirely separate phenomena requiring separate explanations in a fundamental theory. We have already seen that the acceleration equation (6.1) leads not to (6.6) but only to its time derivative (6.3); the omission of the integration constant which gives (6.6) is thus a separate assumption. If, instead, we start from (6.6) and use in addition Maxwell's equation $\operatorname{curl} \mathbf{E} = -\dot{\mathbf{H}}/c$, we find

$$\operatorname{curl}\left(\mathbf{E} - \frac{4\pi\lambda^2}{c^2}\,\dot{\mathbf{J}}\right) = 0, \qquad (6.10)$$

which is, of course, weaker than (6.1) and shows only that

$$\mathbf{E} - \frac{4\pi\lambda^2}{c^2}\,\dot{\mathbf{J}} = \operatorname{grad}\phi, \qquad (6.11)$$

where ϕ is a scalar. For a simply-connected superconductor, there must be no component of grad ϕ in the direction of **J**, for otherwise there would be energy dissipation at the expense of the static magnetic field, which would violate conservation of energy. In this case then, if we ignore any component of grad ϕ perpendicular to **J** we can obtain the acceleration equation (6.1) from the other London equation (6.6). Just here, however, the acceleration equation is of no interest since (6.6) is by itself completely adequate to describe the behaviour of a simply-connected body with no currents led in from outside. It is for a multiply-connected body such as a ring, or for a current fed in from outside, that it is important to prove (6.1), and this turns out to be impossible without further hypotheses. For a ring, for instance, there is nothing to indicate that a non-vanishing value of grad ϕ is inadmissible since ϕ may be a many-valued potential function, though such a value would of course mean that the current in the ring would die away and the flux locked in the ring would disappear. The fact that this does not happen must therefore be regarded as an experimental fact and cannot be deduced mathematically from (6.6).

It will be noticed that there is some suggestion from this discussion that (6.6) is more fundamental than (6.1), since for a simply-connected body (6.1) can be deduced from (6.6) though the reverse is not possible. F. London (1935, 1948, 1950 (see §§ 25, 26)) has, however, shown that (6.6) can be expressed in a rather more significant way which not only provides a clue to what is required of a fundamental theory of superconductivity but also throws some light on the relation between the behaviour of a ring (i.e. zero resistance) and the Meissner effect. In a magnetic field the momentum which determines the de Broglie wave-length is no longer given by $m\mathbf{v}$ but by

$$\mathbf{p} = m\mathbf{v} + e\mathbf{A}/c, \qquad (6.12)$$

where **A** is the vector potential defined so that $\text{div}\,\mathbf{A} = 0$, while the electric current is still determined by **v**. If **p** and **v** refer to local mean values, then bearing in mind the definition of λ, it follows that (6.6) is equivalent to

$$\text{curl}\,\mathbf{p} = 0; \qquad (6.13)$$

and since $\text{div}\,\mathbf{J} = 0$ and $\text{div}\,\mathbf{A} = 0$, we have

$$\text{div}\,\mathbf{p} = 0. \qquad (6.14)$$

For a simply-connected superconductor not fed by a current from outside, the boundary condition $p_n = 0$ applies (since A_n can be made zero, and $J_n = 0$), so the only solution of (6.13) and (6.14) is

$$p = 0. \qquad (6.15)$$

This expresses the fact that the current distribution is uniquely defined by
$$v = -eA/mc. \qquad (6.16)$$

This result is in striking contrast to that for an ordinary conductor where it can be shown that for classical statistical mechanics we must have
$$v = 0. \qquad (6.17)$$

Thus a superconductor is distinguished from a normal conductor in that something prevents the mean momentum p from assuming the local value of eA/c. The mechanism for this prevention is essentially non-classical. Quantum mechanics, however, applied to a free electron gas modifies (6.17) only slightly. In fact, as Landau (1930) has shown, the introduction of quantum mechanics produces only a very feeble diamagnetism instead of the zero magnetism indicated by classical theory. The result (6.15) which summarizes the superconducting behaviour of a simply-connected body requires not only the introduction of quantum mechanics, but also some sort of interaction between electrons, which prevents modification of the electron wave functions when the metal is brought into a magnetic field. A particular kind of interaction which appears to satisfy this requirement has been suggested by Fröhlich (1950) and Bardeen (1950) and will be described briefly in § 6.4.

So far we have discussed the solution of (6.13) and (6.14) only for a simply-connected body with no current fed in from outside. In order to discuss how a persistent current is set up when a flux is put through a ring, F. London uses an ingenious device. He supposes that the ring is not at first superconducting, but embraces a long hollow superconducting cylinder in which a magnetic flux Φ has been frozen. Although there is no field outside this cylinder, we must suppose that a vector potential exists such that round the (non-superconducting) ring

$$\oint (A \cdot ds) = \Phi. \qquad (6.18)$$

Thus, though there is no current in the normal metal, p has the non-zero value eA/c. We now suppose that the ring is made

superconducting—this cannot affect the value of **A** (and therefore
of **p**), since (6.18) must still be satisfied—and we next destroy the
superconductivity of the hollow cylinder. The flux now spreads
outside the hollow cylinder and tries to spread into the material of
the ring; if it is *assumed* that, as in the case of the simply-connected
body, the local value of **p** has a kind of 'rigidity' such that it is unable
to change its value when a magnetic field enters the body, we find
at once that the value of **A** locally is modified by the presence of
magnetic field in the material of the ring. Simple analysis shows, in
fact, that field and current fall off according to the usual penetration
law and that a circulating current flows on the inside rim of the ring,
of just the magnitude given by the macroscopic arguments of
Chapter II. Thus if for simplicity we consider the special case of
a hollow cylinder rather than a ring, then at radius r (6.18) becomes

$$A = \Phi/2\pi r,$$

while after the flux is released from the inner cylinder

$$A = \Phi/2\pi r - \int_x^\infty H\,dx,$$

where x is the depth to which the radius r penetrates into the
cylinder.* If then $mv + eA/c$ is assumed to be unchanged we find,
since v was zero originally, that

$$mv - \frac{e}{c}\int_x^\infty H\,dx = 0,$$

or

$$\frac{dv}{dx} + \frac{eH}{mc} = 0.$$

This is, of course, equivalent to (6.6), and indicates that the current
falls off exponentially from the inside surface of the cylinder. The
current is related to the field H at the boundary which we may take
to be given by $\pi r^2 H = \Phi$, and the relation in this case is simply that
the current per unit length of the hollow cylinder is given by $cH/4\pi$.

The essential point of this discussion is to show that the possibility
of circulating persistent currents in multiply-connected bodies can
be inferred from the single concept of the 'rigidity' of the momentum **p**. London shows that resistanceless flow for the case of current

* Since $r \gg \lambda$, the finite curvature of the cylinder may be ignored in calculating the flux in the metal.

fed into a wire from outside can also be deduced from this concept. This concept thus appears to be more powerful than either of the London equations separately, and is probably an indication that a superconductor with a current round it is to be regarded as a macroscopic quantum mechanism, rather like a giant molecule.

In discussing (6.11) it was pointed out that in order to agree with the experimentally observed zero resistance (and for a simply-connected superconductor to avoid violation of energy conservation) there could be no component of grad ϕ parallel to the current density $\mathbf{J}$. It is not, however, obvious that the component of grad ϕ perpendicular to $\mathbf{J}$ must vanish. On the basis of a relativistic formulation of the theory, it was at first suggested (F. and H. London, 1935 a, b) that ϕ should be identified with $4\pi\lambda^2\rho$, where ρ is the charge density (given by div $\mathbf{E} = 4\pi\rho$). It is easy to show that we should then deduce from (6.11)

$$\nabla^2\mathbf{E} = \mathbf{E}/\lambda^2. \tag{6.19}$$

This would mean that an *electrostatic* field should penetrate into a superconductor to a depth of order λ, in contrast to the almost zero penetration of an ordinary conductor. This question was investigated experimentally by H. London (1936) who looked for a small decrease in the capacity of a condenser when its plates became superconducting, due to this hypothetical increase of penetration of the electrostatic field. No decrease was in fact found, so it could be concluded that grad ϕ in (6.11) is zero.

6.2.3. *Kinetic energy of a superconducting current.* If the energy theorem of Maxwell's theory,

$$\operatorname{div} \frac{c}{4\pi}[\mathbf{E} \wedge \mathbf{H}] + \frac{\partial}{\partial t} \frac{1}{8\pi}(\mathbf{H}^2 + \mathbf{E}^2) = -(\mathbf{J}.\mathbf{E}), \tag{6.20}$$

is applied to a superconductor, the $(\mathbf{J}.\mathbf{E})$ term no longer corresponds to a dissipation of energy, for if we apply (6.1) we see that this term becomes

$$-(\mathbf{J}.\mathbf{E}) = -\frac{\partial}{\partial t}\left(\frac{1}{2}\frac{4\pi\lambda^2}{c^2}\mathbf{J}^2\right),$$

and (6.20) becomes

$$\operatorname{div} \frac{c}{4\pi}[\mathbf{E} \wedge \mathbf{H}] + \frac{\partial}{\partial t}\left(\frac{1}{8\pi}(\mathbf{H}^2 + \mathbf{E}^2) + \frac{1}{2}\frac{4\pi\lambda^2}{c^2}\mathbf{J}^2\right) = 0. \tag{6.21}$$

The term $\dfrac{1}{2}\dfrac{4\pi\lambda^2}{c^2}\mathbf{J}^2$ clearly has the significance of a kinetic energy density of the superconducting current. As we saw in §5.9 this energy is transformed reversibly into the excess of free energy which the normal state has over the superconducting state, when superconductivity is destroyed.

6.2.4. *The size of* λ. According to (6.2) the penetration depth is given by

$$\lambda = (mc^2/4\pi n e^2)^{\frac{1}{2}}. \tag{6.22}$$

It should, however, be emphasized that this is the value indicated by the acceleration theory of Becker, Heller and Sauter (which cannot explain the Meissner effect) and is taken over into the Londons' theory somewhat arbitrarily. The detailed form of the relation can be obtained only in terms of a detailed theory of superconductivity, and we shall see in §6.4 that Fröhlich's theory, for instance, leads to a relation somewhat like (6.22) but differing in detail. Equation (6.22) should really be regarded as little more than a dimensional relation between an 'effective' mass m and an 'effective' number n (or n_s as we shall call it in future) of 'superconducting' electrons per c.c. It is to be expected (though it is not inevitable) that these are of the same order of magnitude as the true electronic mass m_0 and the number n_0 of atoms per c.c. The observed temperature variation of λ will be interpreted in §6.3.2 as due to a temperature variation of n_s, and it is of interest to interpret the experimental values of λ_0 (the penetration depth at $0°$ K.) in terms of n_{s0}, the effective number of 'superconducting' electrons per c.c. at $0°$ K. This is most simply done by calculating the value of the dimensionless parameter r defined by

$$r = (n_{s0}/n_0)/(m/m_0), \tag{6.23}$$

and the values of r shown in Table VII are obtained. Thus for all the elements studied so far r is of order ten times less than the number of valence electrons per atom. This may mean that only

TABLE VII

(See also Table V, p. 150)

Element ...	In	Sn	Hg	Pb
$\lambda_0 \times 10^6$ cm.	6·4	5·2	4·1	3·9
r	0·22	0·35	0·30	0·70
Valency	3	4	2	4

something like a tenth of the electrons become 'superconducting' or that the effective mass is large, but, as already pointed out, such interpretations are probably too naïve. It should not be forgotten, however, that the Londons' theory appeared several years before any experimental data were available, and it is very much to the credit of the theory that it did in fact predict something like the right order of magnitude of λ. In other words, it is satisfactory that r does not differ too drastically from unity; until the detailed theory of superconductivity has been more fully worked out, any more detailed discussion of the precise values of r can only be speculative.

6.2.5. *Generalization of the Londons' theory to anisotropic bodies.* Ginsburg (1944) and von Laue (1947, 1948) pointed out that in the Londons' equation written as

$$c \operatorname{curl}(\Lambda \mathbf{J}) + \mathbf{H} = 0, \tag{6.24}$$

the constant Λ appears as a constant of proportionality between two vectors, so that for an anisotropic body the appropriate generalization is to make Λ a tensor and interpret the equation as

$$c \operatorname{curl}(\mathbf{\Lambda} . \mathbf{J}) + \mathbf{H} = 0, \tag{6.25}$$

where now standard tensor notation is implied. Using this modification it is possible to calculate how the penetration depth (defined by (5.1)) should vary for different orientations of the principal axis* of a single crystal with respect to the direction of current flow, and it is found (Fraser and Shoenberg, 1949) that

$$\lambda = \frac{\lambda_1 \cos^2 \psi + [\lambda_1^2 + (\lambda_3^2 - \lambda_1^2)(1 - \sin^2 \psi \sin^2 \phi)]^{\frac{1}{2}} \sin^2 \psi \cos^2 \phi}{1 - \sin^2 \psi \sin^2 \phi}, \tag{6.26}$$

where ψ is the angle the principal axis makes with the direction of the applied field, and ϕ is the angle the projection of the principal axis on the plane perpendicular to the field makes with the current. Alternatively, the orientation can be described in terms of θ, the angle the principal axis makes with the direction of current flow, and χ the angle the projection of the principal axis on the plane

---

* Only crystals with a single principal axis will be considered (e.g. tin and mercury); the non-vanishing components of the tensor Λ are then $\Lambda_{11} = \Lambda_{22}$, and Λ_{33}.

perpendicular to the current makes with the field. The various angles are indicated in fig. 75, and it is evident that

$$\cos\theta = \sin\psi\cos\phi, \quad \sin\theta\sin\chi = \sin\psi\sin\phi, \quad \sin\theta\cos\chi = \cos\psi,$$
$$(6.27)$$

so that (6.26) can also be expressed as

$$\lambda = \frac{\lambda_1\sin^2\theta\cos^2\chi + [\lambda_1^2 + (\lambda_3^2 - \lambda_1^2)(1 - \sin^2\theta\sin^2\chi)]^{\frac{1}{2}}\cos^2\theta}{1 - \sin^2\theta\sin^2\chi}.$$
$$(6.28)$$

In (6.26) and (6.28), λ_1 and λ_3 are defined as $c(\Lambda_{11}/4\pi)^{\frac{1}{2}}$ and $c(\Lambda_{33}/4\pi)^{\frac{1}{2}}$ respectively, and it follows from (6.26) or (6.28) that they

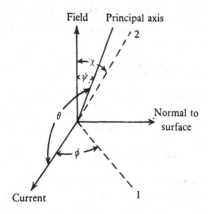

Fig. 75. Illustrating the meanings of θ, ϕ, ψ and χ. The broken lines are the projections of the principal axis on (1) the plane normal to the field, and (2) the plane normal to the current.

are the penetration depths for current flow perpendicular and parallel to the principal axis respectively.

In the experiments of Laurmann and Shoenberg (1949), the specimens were cylinders with the field applied along the axis of the cylinder, so that ψ was constant over the surface of the specimen, but ϕ varied round the circumference, and thus the measured penetration depth λ was an average of (6.26) taken over ϕ. Provided the anisotropy is not too great, this average can be evaluated to a good approximation by expanding (6.26) in powers of α, defined as $(\lambda_1 - \lambda_3)/\lambda_1$, which gives after some reduction

$$\lambda = \lambda_1[1 - \alpha\sin^2\psi\cos^2\phi + \tfrac{1}{2}\alpha^2\sin^4\psi\sin^2\phi\cos^2\phi + \ldots], \quad (6.29)$$

and the average defined as $\bar{\lambda} = \dfrac{2}{\pi} \displaystyle\int_0^{\frac{1}{2}\pi} \lambda \, d\phi$ is given by

$$\bar{\lambda} = \tfrac{1}{2}[\lambda_1 + (\lambda_1 \cos^2 \psi + \lambda_3 \sin^2 \psi)] + \frac{1}{16} \frac{(\lambda_1 - \lambda_3)^2}{\lambda_1} \sin^4 \psi. \quad (6.30)$$

For $\psi = 90°$, this leads to the special form mentioned in §5.4, $\bar{\lambda} = \tfrac{1}{2}(\lambda_1 + \lambda_3)$, if the second term is neglected, and it may be noted that even for such marked anisotropy as λ_1/λ_3 or $\lambda_3/\lambda_1 = \tfrac{1}{2}$, the second term is only 3 % of the first.

In the experiments of Pippard (1950a) the specimens were also cylindrical, but it was the current which had the direction of the cylinder axis, so that the observed value of λ was an average of (6.28) over χ. If (6.28) is expanded in powers of α it becomes

$$\lambda = \lambda_1[1 - \alpha \cos^2 \theta + \tfrac{1}{2}\alpha^2 \cos^2 \theta \sin^2 \theta \sin^2 \chi + \ldots], \quad (6.31)$$

so we see that the variation with χ occurs only in the term proportional to α^2. In other words, if the anisotropy is not too strong λ varies only slightly with χ, and evidently this experimental method is more suited for studying anisotropy than is the Casimir method used by Laurmann and Shoenberg. Taking account of the term in α^2 the average over χ becomes

$$\bar{\lambda} = \lambda_1 \sin^2 \theta + \lambda_3 \cos^2 \theta + \frac{1}{4} \frac{(\lambda_1 - \lambda_3)^2}{\lambda_1} \cos^2 \theta \sin^2 \theta, \quad (6.32)$$

and it is evident that the last term which was omitted in equation (5.15) of §5.4, is of no help in resolving the discrepancy between the theory and the experimental results shown in fig. 66. As has already been pointed out in §5.4, (6.32) is a rather direct consequence of the Londons' theory, so that if further experiments confirm Pippard's result that (6.32) is not valid, the Londons' theory cannot be regarded as better than an approximation. However, it is likely that even if the Londons' equations should prove wrong in detail, they will nevertheless continue to provide a useful qualitative formulation which may prove more tractable mathematically than the more complicated relations which eventually take their place.

6.2.6. *The rotating superconductor.* If a superconductor is rotated it might at first sight seem that since the electrons are 'frictionless' they should be left behind, thus creating a huge circulating current relative to the body. It is easy to see, however, that this will not happen, since as such a current tried to build up it would produce a growing magnetic field, which in turn would produce just the electric field required to accelerate the electrons up to the speed of the body. Because of the electron inertia there should be a slight lagging effect and a rotating superconductor should become slightly magnetized (Barnett effect). The magnitude of this slight effect was calculated by Becker, Heller and Sauter (1933) on the basis of their acceleration theory, and this calculation can be taken over into the Londons' theory, with the one modification that in the Becker, Heller and Sauter calculation the results depend on the initial state, while in the Londons' theory they do not. Thus, according to the acceleration theory, no effect would occur if the body became superconducting when already rotating, while in the Londons' version the body would suddenly acquire the same weak magnetization as if it had been set rotating when already superconducting.

We assume that the acceleration equation

$$m\dot{\mathbf{v}} = e\mathbf{E}, \tag{6.33}$$

is still valid even when the body has a local velocity $\mathbf{v}_0$ ($\mathbf{v}$ is the velocity of the electrons relative to fixed axes). Just as in §6.2.1, this leads to

$$\operatorname{curl}\dot{\mathbf{v}} + \frac{e}{mc}\dot{\mathbf{H}} = 0,$$

and we go over to the Londons' theory by integrating and putting the integration constant equal to zero, i.e.

$$\operatorname{curl}\mathbf{v} + \frac{e}{mc}\mathbf{H} = 0. \tag{6.34}$$

Since, however $\mathbf{J} = ne(\mathbf{v} - \mathbf{v}_0)$, we have

$$\operatorname{curl}\mathbf{H} = \frac{4\pi ne}{c}(\mathbf{v} - \mathbf{v}_0). \tag{6.35}$$

For a rotating body $\mathbf{v}_0 = [\boldsymbol{\omega} \wedge \mathbf{r}],$

where $\boldsymbol{\omega}$ is the angular velocity and $\mathbf{r}$ the radius vector from a point on the axis of rotation, so that

$$\operatorname{curl}\mathbf{v}_0 = 2\boldsymbol{\omega}. \tag{6.36}$$

Combining (6.36) with (6.34) and (6.35) it is easy to show that

$$\nabla^2\mathbf{J} = \mathbf{J}/\lambda^2, \tag{6.37}$$

as before, but that $\nabla^2\mathbf{H} = \dfrac{1}{\lambda^2}\left(\mathbf{H} + \dfrac{2mc}{e}\boldsymbol{\omega}\right). \tag{6.38}$

Thus the current density is, as before, appreciable only within a layer of thickness λ, but it is now not $\mathbf{H}$ which vanishes inside the superconductor but $\mathbf{H} + 2mc\omega/e$. In other words, a very small field

$$\mathbf{H}_0 = -2mc\omega/e \qquad (6.39)$$

is produced inside the superconductor by the surface currents induced by the rotation. Evidently, the current distribution is exactly the same as that which would be produced by an external field $-\mathbf{H}_0$ applied to the stationary body. The external effect of the surface currents is therefore just that of a body with magnetization $\mathbf{I}_0$, where $\mathbf{I}_0$ is the magnetization of the stationary superconductor in an applied field $-\mathbf{H}_0$, and depends on the shape in the manner discussed in Chapter II. Thus for a long rod

$$\mathbf{I}_0 = \mathbf{H}_0/4\pi,$$

and for a sphere $\qquad \mathbf{I}_0 = 3\mathbf{H}_0/8\pi.$

As has already been pointed out in § 2.10.2, the field $\mathbf{H}_0$ given by (6.39), and therefore the apparent magnetization $\mathbf{I}_0$, are very small and have not yet been observed experimentally. The particular interest of an experimental check would be the verification of the uniqueness of the solution given by the Londons' theory—in contrast to the dependence on initial conditions implied by the original Becker, Heller and Sauter approach.

The inverse of the Barnett effect just discussed is the Einstein-de Haas effect—the angular momentum produced by magnetization. As mentioned in § 2.10.2 the Einstein-de Haas effect has already been observed experimentally and the gyromagnetic ratio found to have the value $-e/2mc$ characteristic of electron currents. The mechanism of the effect can be readily understood on the basis of the Becker, Heller and Sauter picture of perfectly free electrons. As the magnetic field is increased, the induced electric field in the penetration layer exerts a torque on the electrons in this layer and an equal and opposite torque on the positive charges of the lattice; since there is no resistive interaction between the electrons and the lattice, the rotation of the lattice persists. The corresponding angular momentum of the lattice is equal but opposite to that of the electrons moving in the superficial current, so the result that the gyromagnetic ratio is $-e/2mc$ follows at once. Since the acceleration equation of

Becker, Heller and Sauter is taken over in the Londons' theory, it is clear that the result is true for that theory also.

A rather general discussion of the problem of the Einstein-de Haas effect has been given by Broer (1947). He points out that the electronic mass m in the gyromagnetic ratio is the real mass, and not an effective mass such as might enter into the formula for the penetration depth. It is to be expected that the same is true also of the Barnett effect, though this is not obvious from the elementary treatment given above.

6.3. The two-fluid model

We have seen in Chapter III that the superconducting transition is one of the 'second kind', but that below the transition temperature the entropy of the superconducting state is always lower than that of the normal state. Thus as the temperature is lowered the degree of order begins to grow as soon as the transition temperature is passed. To provide a more concrete picture of this ordering process, Gorter and Casimir (1934 b) introduced what has come to be known as the 'two-fluid' model.* Originally the model was intended to do no more than reproduce the thermodynamic properties of the metal, but a combination of their theory with the Londons' electrodynamics has given it a much wider application.

6.3.1. *Formulation and thermodynamic properties of the model.* Essentially the idea of the two-fluid model is to suppose that below the transition temperature the metallic electrons are divided between two groups of energy levels. A fraction $(1-x)$ of the electrons occupies the lower group of levels, and these electrons can be loosely described as 'condensed' or 'superconducting'; the remainder (x) remain 'uncondensed' or 'normal'. It is easy to see that the thermodynamic functions cannot depend linearly on x; thus if the Helmholtz free energy f per unit volume is written as

$$f = xf_u(T) + (1-x)f_c(T),$$

then for equilibrium

$$(\partial f/\partial x)_T = 0 \quad \text{or} \quad f_u(T) = f_c(T),$$

which defines a single temperature, presumably the transition temperature, at which x can have values between 0 and 1. At lower

* For a recent discussion see also Gorter (1949).

temperatures $x = 0$ is the only solution, and at higher temperatures $x = 1$ is the only solution. Such a formulation leads in fact to a first-order transition, and indeed this is to be expected from application of the phase rule to two independent phases of a single component. The difficulty can be avoided by supposing that the phases are not independent, and this Gorter and Casimir did by assuming that the thermodynamic functions of the normal phase were not proportional to x. More specifically, they assumed the free energy* per unit volume to be

$$f = x^r f_u(T) + (1-x) f_c(T), \tag{6.40}$$

where
$$f_u = -\tfrac{1}{2} a T^2, \quad f_c = -\beta. \tag{6.41}$$

The choice of the function x^r is, of course, arbitrary, but can be fitted to the experimental facts if the further assumption $r = \tfrac{1}{2}$ is made. In order to simplify the presentation, we shall omit the proof that $r = \tfrac{1}{2}$ is the best assumption and develop the consequences of the more special form

$$f = x^{\frac{1}{2}} f_u(T) + (1-x) f_c(T). \tag{6.42}$$

The form of the functions f_u and f_c is chosen to agree with the linear temperature dependence of entropy of a 'normal' electron gas, and with the absence of any entropy transport by the 'superconducting' electrons, i.e. the absence of Thomson heat (see § 3.7).

The equilibrium value of x at any temperature is given by $(\partial f/\partial x)_T = 0$ or
$$-\beta + a T^2 / 4 x^{\frac{1}{2}} = 0,$$
so that
$$x = (a T^2 / 4\beta)^2.$$
Now for $T = T_c$, we must have $x = 1$, so

$$\beta = a T_c^2 / 4. \tag{6.43}$$

and we have
$$x = (T/T_c)^4. \tag{6.44}$$

Equation (6.42) now becomes

$$f = -\tfrac{1}{4} a T_c^2 - \tfrac{1}{4} a T^4 / T_c^2, \tag{6.45}$$

the entropy per unit volume is

$$s = -df/dT = a T^3 / T_c^2, \tag{6.46}$$

and the specific heat per unit volume is

$$c = 3 a T^3 / T_c^2. \tag{6.47}$$

---

* The free energy discussed here refers only to the electrons, and it is important to bear this in mind when comparing the results of the analysis with the discussion of § 3.2.

Comparing these results with (3.6) and (3.6a) (which describe the experimental behaviour in a slightly idealised way) and bearing in mind that (6.47) ignores the lattice specific heat (the bT^3 term of (3.6)), we see that the two-fluid model does so far correctly reproduce the experimental facts. This is, of course, not surprising, since the assumptions have been carefully chosen to do just this. It is, of course, possible to derive also the critical field curve, and again get the correct answer (3.8), but this is no merit of the model since, as we saw in § 3.2, this parabolic law (3.8) is a direct consequence of assuming a cubic temperature variation of specific heat in the superconducting state.

In terms of this model, the reason why the specific heat curve in the superconducting state crosses the normal curve and rises above it, is that extra heat has to be supplied to evaporate condensed electrons into the 'normal' phase as the temperature is raised. As soon, however, as the transition temperature is reached all the electrons have been evaporated ($x = 1$), and this extra contribution disappears abruptly; the specific heat then drops to the value characteristic of the 'normal' phase alone.

It is convenient to note here also that the internal energy per unit volume which is given by

$$u = f - T \, df/dT$$

becomes
$$u = -\tfrac{1}{4}aT_c^2 + \tfrac{3}{4}aT^4/T_c^2. \qquad (6.48)$$

From (6.43) and (3.7) we see that for $T = 0$, the internal energy can be expressed as

$$u(0) = -\tfrac{1}{4}aT_c^2 = -\beta = -H_0^2/8\pi. \qquad (6.49)$$

The idea of an energy gap β between the 'condensed' and 'uncondensed' states was, of course, deliberately introduced in the initial formulation (6.41).

6.3.2. *Combination with the Londons' theory.* So far the two-fluid model has not given anything new beyond providing a useful picture of the superconducting transition as a 'condensation' process. Its main success—apart from some qualitative applications which we shall discuss later—is when it is combined with the Londons' theory. If it is supposed that the proportion of 'superconducting' electrons in the Londons' theory is equal to $(1 - x)$, the fraction of electrons condensed in the Gorter-Casimir theory, we find from (6.22) and (6.44) that

$$\left(\frac{\lambda}{\lambda_0}\right)^2 = \frac{1}{(1-(T/T_c)^4)}, \qquad (6.50)*$$

which, as we saw in §5.2.1 (see equation (5.10)), is just the relation found by experiment,* and the value of λ_0 is given by

$$\lambda_0 = (mc^2/4\pi n e^2)^{\frac{1}{2}}, \qquad (6.51)$$

where now n means the total effective number of free electrons per unit volume in the metal (since $x=0$ at $T=0$, *all* these electrons must have condensed at absolute zero).

The presence of uncondensed or 'normal' electrons requires a modification of the Londons' electrodynamic equations when non-stationary processes are considered. If, in fact, an electric field is present it will not only accelerate the superconducting electrons, but also cause a current of normal electrons. The simplest assumption to make about the normal electrons is that they obey Ohm's law, i.e. that

$$\mathbf{J}_n = \sigma \mathbf{E}. \qquad (6.52)$$

The equations of §6.2 now refer to only the current density $\mathbf{J}_s$ of superconducting electrons and (6.6) and (6.1) now become

$$\mathbf{H} = -\frac{4\pi\lambda^2}{c}\,\mathrm{curl}\,\mathbf{J}_s, \qquad (6.53)$$

$$\mathbf{E} = \frac{4\pi\lambda^2}{c^2}\,\dot{\mathbf{J}}_s. \qquad (6.54)$$

The total current $\mathbf{J}$ which enters into Maxwell's equations is given by

$$\mathbf{J} = \mathbf{J}_n + \mathbf{J}_s. \qquad (6.55)$$

As we shall see below, these modifications are important only for very rapidly changing fields, and do not affect any of the results we have already obtained. Their main application is, in fact, in the theory of the high-frequency behaviour of superconductors.

6.3.3. *The high-frequency behaviour of superconductors.* The first theoretical treatment of this problem was given by H. London (1934, 1940) and provides a convenient starting point. We shall see

---

* This formula was first brought to my notice by Professor J. G. Daunt in a private conversation in 1946; see also Daunt (1947) and Daunt, Miller, Pippard and Shoenberg (1948).

that because of the anomalous properties of normal metals at high frequencies and low temperatures, this theory proves to be inadequate, but because the detailed theory of these anomalous properties (Pippard, 1947b; Reuter and Sondheimer, 1948) is rather complicated, we shall do no more than outline the salient points of the much more detailed discussion in which Pippard (1947b,c, 1950a,b) has taken account of these anomalous properties.

H. London applied the equations of the previous section to the case where all fields are proportional to $e^{i\omega t}$. This leads immediately to

$$J = \left(\sigma - \frac{ic^2}{4\pi\lambda^2\omega}\right) E, \qquad (6.56)$$

and this equation now replaces Ohm's law in the working out of the usual theory of the skin effect. The results are most simply discussed in terms of the surface impedance, which is defined as

$$Z = E_0 \bigg/ \int_0^\infty J\, dx. \qquad (6.57)$$

It is here assumed that the surface can be treated as a plane (which is evidently permissible provided the fields do not penetrate deeply), and x is measured along the normal to the surface, with $x=0$ at the surface; E_0 is the value of E at $x=0$. This surface impedance is the complex equivalent of the resistance per unit square of surface; its real part is the actual resistance R per unit square and its imaginary part is the reactance X.

A straightforward calculation in which (6.56) is combined with Maxwell's equations shows that

$$H = H_0 e^{-kx}, \quad E = E_0 e^{-kx}, \quad J = J_0 e^{-kx}, \qquad (6.58)$$

where
$$k = \frac{1}{\lambda\sqrt{2}}\left((m+1)^{\frac{1}{2}} + i(m-1)^{\frac{1}{2}}\right), \qquad (6.59)$$

$$m = (1 + 4\lambda^4/\delta^4)^{\frac{1}{2}} \qquad (6.60)$$

and
$$\delta = c/(2\pi\omega\sigma)^{\frac{1}{2}}. \qquad (6.61)$$

The physical meaning of δ is, of course, just the skin depth that would be characteristic of a normal metal of conductivity σ, and (6.59) shows that the presence of superconducting electrons complicates the decay law. It should be noticed that in the limiting cases $\lambda/\delta \ll 1$ (i.e. low frequencies) and $\lambda/\delta \gg 1$ (high frequencies*), we find

$$k = 1/\lambda \quad \text{and} \quad k = (1+i)/\delta \qquad (6.62)$$

respectively. These correspond, as they should, to ordinary super-conducting penetration for static or slowly varying fields, and to the ordinary skin effect behaviour for such high frequencies that the superconducting electrons no longer matter. Indeed, it is already evident from (6.56) that the characteristic effects of high frequencies will be important only at frequencies higher than that for which

$$4\pi\lambda^2\omega/c^2 \sim 1/\sigma$$

or

$$\lambda \sim \delta.$$

If we put for σ the value characteristic of a very pure metal such as tin in the normal state, i.e. $\sigma \sim 5 \times 10^{20}$ e.s.u., and $\lambda \sim 5 \times 10^{-6}$ cm., this gives ω of order 6×10^9 sec.$^{-1}$ or a frequency of 10^3 mcyc./sec. as that at which 'characteristic' effects may be expected to set in. This is just the order of the frequency used in London's and Pippard's experiments (see § 1.4.6).

If (6.58) to (6.61) are now substituted into (6.57), and (6.56) is used, the surface impedance is found after some reduction to be

$$Z = \frac{4\pi i\omega\lambda}{mc^2\sqrt{2}}((m+1)^{\frac{1}{2}} - i(m-1)^{\frac{1}{2}}), \qquad (6.63)$$

and thus

$$R = 4\pi\omega\lambda(m-1)^{\frac{1}{2}}/mc^2\sqrt{2}, \qquad (6.64)$$

and

$$X = 4\pi\omega\lambda(m+1)^{\frac{1}{2}}/mc^2\sqrt{2}. \qquad (6.65)$$

For the normal metal $\qquad R_n = 2\pi\omega\delta_n/c^2, \qquad (6.66)\dagger$

where δ_n is the value of δ obtained by substituting σ_n, the value of the normal conductivity, for σ in (6.61). So

$$\frac{R}{R_n} = \frac{\lambda\sqrt{2}}{\delta_n}\frac{(m-1)^{\frac{1}{2}}}{m}. \qquad (6.67)$$

Except very close to the transition temperature T_c, we can assume that owing to the decrease of the number of 'normal' electrons below T_c, σ is appreciably smaller than σ_n, and hence δ is appreciably

---

* According to standard theory, relaxation effects for the normal electrons would become important at frequencies as high as this. Ohm's law should in fact be replaced by $J + \tau \dot{J} = \sigma E$, where τ is the relaxation time, for frequencies such that $\omega\tau \gtrsim 1$; since $\sigma = ne^2\tau/mc^2$ this condition is equivalent as regards order of magnitude to $\delta \lesssim \lambda$. Actually, however, when the anomalous skin effect is considered it turns out that relaxation effects become important only at much higher frequencies.

† This result is easily obtained from (6.64) by considering the limiting case $m \gg 1$.

greater than its normal value δ_n. Since λ also falls with temperature, we can assume that $4\lambda^4/\delta^4$ is small and (6.67) then reduces to

$$R/R_n = 2\lambda^3/\delta_n\delta^2. \tag{6.68}$$

Thus fairly close to T_c, R/R_n should drop steeply both because of the rapid fall of λ, and of the growth of δ (due to a falling conductivity of the normal electrons). At lower temperatures the fall should become more gradual since λ is then practically constant. Finally, at absolute zero there should be no 'normal' electrons left at all and R should vanish. Evidently also at any given temperature R/R_n should increase with frequency on account of (6.61). As can be seen from fig. 4 of § 1.4.6, all these predictions are in excellent qualitative agreement with the facts, but difficulties arise immediately any quantitative comparison is attempted.

It is, in fact, necessary to make some specific assumption about the behaviour of σ, the conductivity of the normal electrons, and in the absence of any detailed theory any such assumption is of course little more than guesswork. H. London (1940) chose the simplest assumption that σ is $x\sigma_n$ where x is the normal fraction of the electrons, as defined in § 6.3.1. This, if taken together with equation (6.44) for x, gives

$$\frac{R}{R_n} = 2\left(\frac{\lambda_0}{\delta_n}\right)^3 \frac{(T/T_c)^4}{(1-(T/T_c)^4)^{\frac{3}{2}}}, \tag{6.69}$$

for temperatures such that $4\lambda^4/\delta^4 \ll 1$, i.e. for

$$T/T_c \lesssim (1 + 2\lambda_0^2/\delta_n^2)^{-\frac{1}{4}}. \tag{6.70}$$

In the conditions of H. London's experiment on tin,

$$\omega = 2\pi \times 1.46 \times 10^9 \text{ sec.}^{-1}$$

and $\sigma_n = 1.35 \times 10^{20}$ e.s.u., so $\delta_n = 10.7 \times 10^{-6}$ cm.; thus putting $\lambda_0 = 5.2 \times 10^{-6}$ cm., we see that (6.69) should hold for $T/T_c \lesssim 0.9$. If λ_0/δ_n is regarded as an adjustable parameter, (6.69) can be fitted fairly well to the experimental results (though, as Pippard (1947c) has shown, even the shape of the curve is not quite right), but if the above value of λ_0/δ_n is taken, the calculated values of R/R_n come out appreciably greater than the observed ones. Thus, using the data appropriate to London's experiment,* (6.69) predicts that for $T/T_c = 0.7$, $R/R_n \sim 0.1$, while the observed value is about 4×10^{-3}.

* At the time of London's experiment the absolute value of λ_0 was not known and London made a choice to fit his results. He assumed also a much smaller value of σ_n to take account of the 'anomalous skin effect' (see next paragraph).

Apart from this numerical discrepancy, a more serious failing of (6.69) is that it predicts that R/R_n should vary as $\omega^{\frac{2}{3}}$, i.e. should increase by a factor of 22 in going from 1200 mcyc./sec. to 9400 mcyc./ sec., while Pippard found a factor of only between 4 and 9.*

The clue to these discrepancies was provided by the observation that even the prediction for the normal metal was not fulfilled. Thus London found that the absolute value of R_n was nearly three times higher than it should be according to (6.66). He suggested that this discrepancy might be due to the fact that at low temperatures the electronic mean free path l was quite large compared to the classical skin depth δ_n. In these circumstances the electric field accelerating an electron may vary considerably over a free path, and Ohm's law must be replaced by a relation in which the current density depends on space integrals of the electric field. The theory of this 'anomalous skin effect' was worked out in a semi-quantitative form by Pippard (1947 b) and in a more detailed mathematical form by Reuter and Sondheimer (1948), and has proved to give very satisfactory agreement with experiments by Pippard (1947 a, 1950 a) and Chambers (1950, and to be published) in its predictions of the variation of R_n both with σ_n and ω.

Clearly, if Ohm's law has to be modified for the electrons in the normal metal, it must also be modified for the 'normal' electrons in the superconducting state, i.e. (6.52) must be replaced by a more complicated relation. The consequences of this modification have been worked out in detail by Maxwell, Marcus and Slater (1949), but since the salient features of their theory may be reproduced by simpler arguments we shall not discuss the detailed calculations. Pippard (1947 b) showed that the 'anomalous skin effect' in normal metals could be fairly well represented by the 'ineffectiveness concept', according to which the ordinary theory of the skin effect can be used, if σ_n is replaced in the 'anomalous' region by

$$\sigma_n' = \beta \delta_n \sigma_n / l. \tag{6.71}$$

Here l is the mean free path, β a numerical factor of order unity, and δ_n the skin depth, which is now given by the implicit relation

$$\delta_n = c/(2\pi\omega\sigma_n')^{\frac{1}{2}}, \tag{6.72}$$

* The comparison was not made on the same specimen, and since the detailed results varied considerably with crystal orientation, it is possible only to set these limits to the factor.

or, substituting for σ_n' from (6.71), by

$$\delta_n = (c^2 l/2\pi\beta\omega\sigma_n)^{\frac{1}{3}}. \qquad (6.73)$$

Since σ_n is proportional to l we get the remarkable result that δ_n and hence R_n ($= 2\pi\omega\delta_n/c^2$) is independent of σ_n, the d.c. conductivity. This is in good agreement with the experimental results for sufficiently high d.c. conductivities. The physical justification for the assumption (6.71) is that if $l \gg \delta$ only those electrons moving at angles smaller than about δ/l travel in the electric field for any appreciable fraction of their mean free path; thus the conductivity is effectively reduced by a factor of order δ/l.

The application of this approximate concept to the superconducting metal suggests* that equations (6.63)–(6.68) will still be valid if, instead of σ, we substitute σ', given by

$$\sigma' = \beta x \lambda \sigma_n/l, \qquad (6.74)$$

since, except rather close to the transition temperature, the field penetrates to a depth which is approximately λ (as already mentioned, the real part of k in (6.59) is very nearly $1/\lambda$, provided $\lambda \ll \delta$). Thus we now have

$$\delta = c/(2\pi\omega\sigma')^{\frac{1}{2}}. \qquad (6.75)$$

Combination of (6.72) and (6.75), together with (6.71) and (6.74), gives

$$\frac{\delta^2}{\delta_n^2} = \frac{\sigma_n'}{\sigma'} = \frac{\delta_n}{x\lambda}, \qquad (6.76)$$

and the approximation (6.68), which is valid except close to T_c, gives

$$R/R_n = 2x(\lambda/\delta_n)^4, \qquad (6.77)$$

which becomes

$$\frac{R}{R_n} = 2\left(\frac{\lambda_0}{\delta_n}\right)^4 \frac{(T/T_c)^4}{(1-(T/T_c)^4)^2}. \qquad (6.78)$$

The main difference between this result and (6.69) is that δ_n now has a different meaning (see (6.73)); the temperature variation is not very seriously different and agrees about as well with the experimental data as did (6.69) (see equation (6.86). The anomalous value of δ_n in London's experiment is most simply deduced from the value of R_n found experimentally; this gives a value 2·7 times the 'classical' value of p. 200, i.e. $\delta_n = 2\cdot9 \times 10^{-5}$ cm. The value of R/R_n from (6.78) at $T/T_c = 0\cdot7$ thus becomes $0\cdot9 \times 10^{-3}$, which though smaller than the experimental value (4×10^{-3}), is

---

* The presentation given below is a simplified and therefore more approximate version of that due to Pippard (1947c).

a good deal nearer to it than was the value 0·1 predicted by (6.69). The frequency variation predicted by (6.78) is that R/R_n should vary as $\omega^{\frac{4}{3}}$; thus, in going from 1200 to 9400 mcyc./sec., the value of R/R_n should increase by a factor 15 which is still greater than the observed factor of between 4 and 9, but nearer than the factor 22 given by an $\omega^{\frac{3}{2}}$ variation.

Although it is not surprising that the simplified treatment presented above should not give a completely satisfactory interpretation of the experiments, it turns out that even the more detailed analysis reproduces some of the discrepancies. Thus the prediction that R/R_n at low temperatures should vary as $\omega^{\frac{4}{3}}$ comes also out of the more detailed analysis, and, indeed, Pippard (1950 b) has pointed out that this variation is a direct consequence of any simple two-fluid model. If we assume that the essential characteristic of any two-fluid model is that it is possible to divide the current into two independent parts, then it is clear that at low temperatures the greater part of the current is carried by the superconducting electrons. In these conditions the configuration of electric and magnetic fields in the skin layer will be determined mainly by the 'supercurrent' (this is confirmed by the results obtained for the reactance X which will be discussed below). Thus, for a given total current, the magnitude and configuration of the magnetic field should be independent of frequency, so that the electric field which can be considered as induced by the oscillating magnetic field should have a fixed configuration but a magnitude proportional to ω. Now the electric field causes a current proportional to itself and therefore a power dissipation proportional to ω^2, so that the value of R should vary as ω^2. Since, however, R_n varies as $\omega^{\frac{2}{3}}$ (see (6.66) and (6.73)), it follows that R/R_n should vary as $\omega^{\frac{4}{3}}$.

In order to compare theory and experiment in the region close to the transition temperature (which we have not so far discussed), it is convenient first to develop a further dimensional argument due to Pippard (1950 b), which also leads to (6.77), but avoids some of the special assumptions previously made. Whatever the relation between J_n and E, it is possible to define a skin depth δ'_n which is such that the surface resistance R'_n of a fictitious metal containing only the 'normal' electrons would be given by

$$R'_n = 2\pi\omega\delta'_n/c^2. \qquad (6.79)$$

Since the only difference between such a fictitious metal and the actual normal metal is that the d.c. conductivity of the fictitious metal is x times smaller owing to the reduction in the number of electrons, and since the skin depth δ_n in the normal metal is proportional to $(\omega\sigma/l)^{-\frac{1}{3}}$ (see (6.73)), it is plausible to assume*

$$\delta'_n = x^{-\frac{1}{3}}\delta_n, \text{ i.e. } R'_n = x^{-\frac{1}{3}}R_n. \qquad (6.80)$$

If R is the actual resistance of the superconductor, it is reasonable on dimensional grounds to suppose that

$$R/R'_n = \Phi(\lambda/\delta'_n), \qquad (6.81)$$

where Φ is some unspecified function, since λ is the only new parameter which enters into the equation when we go over from the fictitious metal to the actual superconductor. Thus finally

$$R/R_n = x^{-\frac{1}{3}}\Phi(u), \qquad (6.82)$$

where $$u = 2\pi\omega x^{\frac{1}{3}}\lambda/R_n c^2. \qquad (6.83)$$

If now the previous electrodynamic argument is also used, which shows that at low temperatures R/R_n should vary as $\omega^{\frac{4}{3}}$, we see that since R_n varies as $\omega^{\frac{2}{3}}$, Φ must be a quartic function for low values of u, so that $$R/R_n \propto x(\lambda/\delta_n)^4. \qquad (6.84)$$

If x is taken to have the same meaning as the x of the Gorter-Casimir theory, then this is in agreement with the more special argument which led to (6.77).

Equation (6.82) makes interesting predictions as to how R/R_n should change from one specimen to another at any particular temperature and frequency. Specimens may differ from each other both as regards purity (i.e. d.c. conductivity in the normal state) and crystal orientation. As regards the influence of purity, R_n should be independent of σ_n, because extreme 'anomalous' conditions apply, and this has recently been verified by Pippard (to be published) in experiments on tin contaminated with indium;† λ, too, was found not to change appreciably, as might be expected, since

---

* In this argument, x should be regarded as the reduction factor of the d.c. conductivity rather than as the fraction of normal electrons. Previously we have assumed that the two definitions were identical, but this is not necessary here. Note, too, that since σ is proportional to l, and only σ/l is involved, any change of l due to the superconductivity of the metal will not matter, since σ will change in proportion.

† A 20-fold increase of the d.c. resistivity had no appreciable effect on R_n, but for greater contaminations, the mean free path was so much reduced that some variation was observed (in agreement with the Reuter-Sondheimer theory).

the transition temperature and thermodynamic properties are not appreciably affected by contamination. Thus u is independent of contamination, and according to (6.82) this should mean that R/R_n as a function of temperature and frequency should also be independent of contamination; this again has been verified experimentally by Pippard.

The influence of crystal orientation has also been studied by Pippard (1950a), who found that R_n and λ are both affected (for a discussion of the influence on λ see §5.4). Since

$$\lambda = \lambda_0/(1-(T/T_c)^4)^{\frac{1}{2}}$$

and R_n is independent of temperature, while all the other parameters entering into u may be assumed to be independent of orientation and to depend only on temperature (at a given frequency), (6.82) may be expressed as

$$R/R_n = G_2 f(T), \tag{6.85}$$

where f is the same function for all specimens and G_2 is a scaling factor depending only on the crystal orientation of the specimen.* This is exactly what Pippard finds experimentally at 9400 mcyc./sec. but with $f(T)$ to a good approximation given by the empirically determined function

$$f(T) = (T/T_c)^4 (1-(T/T_c)^2)/(1-(T/T_c)^4)^2, \tag{6.86}$$

rather than by (6.78), which would hold if Φ were a quartic function and x had the same meaning as in the Gorter-Casimir theory, i.e. if x were given by $(T/T_c)^4$.

We now consider what form is assumed by (6.82) close to the transition temperature. Here the dominating temperature variation arises from λ which, in accordance with (5.11), we can put equal to $\frac{1}{2}\lambda_0 T_c^{\frac{1}{2}}(T_c - T)^{-\frac{1}{2}}$, and x can be treated as constant (equal to unity).

Thus

$$\frac{R}{R_n} = \Phi\left(\frac{\pi\omega\lambda_0 T_c^{\frac{1}{2}}}{c^2 R_n (T_c - T)^{\frac{1}{2}}}\right), \tag{6.87}$$

or

$$\frac{R}{R_n} = \psi\left(\frac{G_1}{T_c - T}\right), \tag{6.88}$$

where G_1 is a scaling factor depending only on the crystal orientation of the specimen. This again is just what is found experimentally, i.e. in the neighbourhood of T_c the experimental curves, if plotted against $T_c - T$, can be brought into coincidence by merely changing

* This is true only if Φ is such that $\Phi(u_1 u_2) = \Phi(u_1)\Phi(u_2)$, i.e. $\Phi(u) \propto u^r$.

the scale of $(T_c - T)$. When, however, the values of G_1 and G_2 are compared, the theory fails. Thus from the definitions of G_1 and G_2 we should have

$$G_2 \propto \Phi(G_1^{\frac{1}{2}}), \qquad (6.89)$$

and since Φ should be a quartic, we should have $G_2 \propto G_1^2$. Actually, however, the experiments indicate with fair accuracy that $G_1 \propto G_2$, thus providing yet another inconsistency of the two-fluid model.

These various discrepancies all suggest that the two-fluid model does not fit the facts more than qualitatively in its present form. Whether it can be retained at all is an open question. Perhaps it is not too much to hope that the fault lies not so much in the two-fluid model itself as in the mode of its application to this complicated problem. For instance, it is possible that the assumption of complete independence of the super and normal currents may have to be abandoned.

Up to now the discussion has been limited mainly to the resistive part R of the surface impedance Z, and we shall now discuss briefly the predictions of the theory regarding the reactive part X. According to (6.65), the surface reactance when $\lambda \ll \delta$ is just

$$X = 4\pi\omega\lambda/c^2, \qquad (6.90)$$

since then m is very nearly unity. This means that the field configuration at low temperatures is determined almost entirely by the static penetration depth, independently of the frequency. This was, in fact, assumed in the dimensional argument which showed that R/R_n should vary as $\omega^{\frac{4}{3}}$. We have already described in §5.3.3 how Pippard was able to measure not the effective penetration depth itself but changes in it with temperature, by studying the changes in resonant frequency when superconductivity was destroyed by a magnetic field. The 'effective' penetration depth λ' introduced in §5.3.3 is really defined by

$$X = 4\pi\omega\lambda'/c^2, \qquad (6.91)$$

and Pippard's results are consistent with the prediction that for sufficiently low temperatures λ' is identical with λ. At higher temperatures λ' falls below λ (see fig. 65b), and this again is predicted theoretically by (6.65), since as T_c is approached m becomes appreciably greater than unity. Pippard (1950b) has shown by a dimensional argument that we should expect λ'/λ to be given by a function of δ_s/λ (where δ_s is defined by $R = 2\pi\omega\delta_s/c^2$). The form of this function can be calculated from the detailed theory, and though qualitative agreement is obtained, there are again discrepancies in detail. Perhaps the most disturbing of these is that the experiments indicate significantly different functions for different crystal orientations.

6.3.4. *The high-frequency resistance at absolute zero.* We have seen that up to 9400 mcyc./sec. there is no evidence for any resistance at absolute zero, in agreement with the assumption of the two-fluid model that there all the electrons have 'condensed'. At still higher frequencies we might, however, expect quantum processes to set in, which could raise electrons from the condensed to the uncondensed state and thus cause energy absorption.

At first sight it might seem that the frequency for which quantum transitions become possible is simply given by

$$h\nu = \beta/n = aT_c^2/4n = H_0^2/8\pi n,$$

since β/n is the energy difference per electron between the condensed and the uncondensed states at absolute zero. Since

$$a/n \sim k/T_0,$$

where T_0 is the degeneracy temperature of the electrons in the normal metal (of order 10^4 °K.), this gives

$$h\nu \sim kT_c^2/T_0. \qquad (6.92)$$

Putting in numerical values, we find $\nu \sim 10^7$ cyc./sec., which is much lower than indicated by experiment. Evidently this interpretation is too naïve, and the absorption of energy must be a collective process involving many electrons. The frequency given by

$$h\nu \sim kT_c, \qquad (6.93)$$

i.e. $\nu \sim 10^{11}$ cyc./sec., is about ten times higher than Pippard's highest frequency and probably much nearer the true critical frequency. As mentioned in § 1.4.6, this frequency is unfortunately in a region which is technically awkward.

6.3.5. *Dependence of penetration depth on field; long-range order.* *
In the evaluation of the fraction x of 'normal' electrons by the Gorter-Casimir theory (§ 6.3.1) the Helmholtz free energy was

---

* The theory described in this section was developed by Pippard (1950 c, 1951 a) following a suggestion by H. London (unpublished) that since the penetration depth varies with temperature, the entropy must vary with magnetic field (to satisfy the thermodynamic relation $\partial S/\partial H = \partial I/\partial T$). This field dependence of entropy must presumably arise from a field dependence of x, which in principle can be derived by an extension of the Gorter-Casimir theory.

minimized, since it was assumed that there was no applied magnetic field. If, however, x and hence λ depend on the applied field, the correct value of x is that which minimizes the Gibbs free energy. The rigorous solution of the problem if x is taken as varying continuously in the penetration region is, however, very complicated, since the Gibbs free energy cannot be explicitly written down until the field distribution is known, while the field distribution will itself depend on the variation of x. Indeed, if λ is a function of field, the Londons' equation is no longer linear, so that the usual solutions can at best be only approximately valid. Fortunately, however, the problem can be discussed by a much simpler argument, due to Pippard (1950c), which both shows that the rigorous argument is bound to give a field dependence of λ much stronger than actually observed (see § 5.5) and indicates a sounder interpretation of the facts. Pippard considers unit area of a macroscopic slab of superconductor, and supposes that to a depth a, which provisionally is assumed a good deal greater than λ, the value of x is constant but is a function of the field H at the surface, as well as of temperature. The Gibbs free energy can be written as

$$G = 2G_1 + G_2,$$

where G_1 is the Gibbs free energy of each of the laminae of thickness a at the two surfaces of the slab, and G_2 is the free energy of the rest of the slab. Since $a \gg \lambda$, the field is zero inside the rest of the slab and so x has the value x_0 characteristic of zero field, and G_2 is the same as the Helmholtz free-energy function. Since G_1 and G_2 are functions of the different variables x and x_0, they may be separately minimized; minimization of G_2 leads to the result previously obtained:

$$x_0 = (T/T_c)^4,$$

but the minimization of G_1 must be considered separately. Bearing in mind the results of § 6.3.1 ((6.41), (6.42) and (6.49)*), we can write

$$G_1 = -\frac{aH_0^2}{8\pi}\left((1-x) + 2x^{\frac{1}{2}}\left(\frac{T}{T_c}\right)^2\right) + (a-\lambda)\frac{H^2}{8\pi}.$$

We have also that $\lambda = \lambda_0/(1-x)^{\frac{1}{2}}$, so on minimizing with respect to x we find

$$1 - \left(\frac{x_0}{x}\right)^{\frac{1}{2}} - \frac{\lambda_0}{2a(1-x)^{\frac{3}{2}}}\left(\frac{H}{H_0}\right)^2 = 0,$$

---

* The a of § 6.3.1, which is equal to $H_0^2/2\pi T_c^2$, should not be confused with the depth a of this section.

which leads to a change of penetration depth in field H given approximately by

$$\frac{\Delta\lambda}{\lambda} = \frac{(\lambda_0/a)(H/H_0)^2}{2(1-x_0)^{\frac{3}{2}}/(x_0) - 3(\lambda_0/a)(H/H_0)^2}. \qquad (6.94)$$

A typical experimental value of $\Delta\lambda/\lambda$ found by Pippard was 2×10^{-2} at $3\cdot6°$ K. when the applied field was critical; substituting numerical values $(H/H_0 = 0\cdot064, x_0 = 0\cdot88)$, this gives $a \sim 20\lambda_0$; values of a deduced at lower temperatures are somewhat larger. This result justifies the basis of the treatment and also the approximations used, for it is clear that if x had been regarded as *continuously* varying with field, this method of approach would be a valid simplification only if a was of the same order of magnitude as λ. The conclusion is then that there must be a long-range order in the superconductor which prevents the degree of condensation $(1 - x)$ changing except over fairly large distances—of order $20\lambda_0$, or 10^{-4} cm., in pure tin. This distance is much greater than the size of the smallest specimen exhibiting superconducting properties, for we saw in Chapter V that films only 5×10^{-7} cm. thick, and colloid particles of radius of order only 10^{-6} cm. still had the same transition temperature as bulk material. The range of order must therefore not be regarded as a minimum range necessary for the setting up of the ordered state, but rather as the range to which order will extend in the bulk material. We may picture ordered regions in the super-conductor growing in size until further growth is inhibited by impurities or thermal vibrations, or, in small specimens, by the boundaries of the material. According to these ideas, the range of order should get smaller in impure specimens, so that the dependence of penetration depth on field should be more marked in less pure specimens; this has been recently confirmed by Pippard (1951 b).

For thin specimens the application of the Gorter-Casimir theory becomes straightforward once it can be taken for granted that x is constant throughout the specimen (Lock, 1950).* Thus the Gibbs free energy per unit volume of a plate of thickness $2a$ can be written as

$$g = -\frac{H_0^2}{8\pi}\left((1-x) + 2x^{\frac{1}{2}}\left(\frac{T}{T_c}\right)^2\right) + \frac{H^2}{8\pi}\left(1 - \frac{\lambda}{a}\tanh\frac{a}{\lambda}\right), \qquad (6.95)$$

---

* The discussion ignores any surface energy term, but the considerations of § 5.8 suggest that this is justified.

(see Appendix II, equation (A 2)). This can now be minimized with respect to x, bearing in mind that $\lambda = \lambda_0/(1-x)^{\frac{1}{2}}$, thus giving an equation for H as a function of x. Since the magnetization I is also a function of λ and so of x (see Appendix II, equation (A 2)), the magnetization curve can be constructed by substituting different values of x into the equations for I and H. The curves calculated by Lock in this way do in fact reproduce the qualitative features of the experimental results described in § 5.5 (see fig. 67), but the experimental curves show a more marked rounding than the theoretical ones. It should be noticed that at very low temperatures, x must always approach zero, so that whatever the field dependence of x comes out to be, λ must stay close to λ_0 at all fields up to the critical field. Thus an essential feature of the predictions of this type of theory is that penetration depth should be independent of field at sufficiently low temperatures, even for very thin films. Lock's magnetization curves of thin films do show some straightening out of the magnetization curves as the temperature is lowered, but it is unlikely that they become completely straight even at the lowest temperatures; other reasons for rounding of the magnetization curves have been mentioned in §§ 5.5 and 5.8. Similarly, the theory outlined above cannot explain the abnormally high critical fields of thin films in the experiments of Appleyard *et al.* and of Shalnikov and Alekseyevsky, since it is clear that their critical field values even at absolute zero exceed the thermodynamic values (5.25) based on a linear magnetization curve by more than could be attributed to the effect of the surface energy parameter β.

A somewhat similar analysis (Pippard, 1952) may also be made for small spheres, with (6.95) replaced by

$$g = -\frac{H_0^2}{8\pi}\left(1-x+2x^{\frac{1}{2}}\left(\frac{T}{T_c}\right)^2\right) + \frac{3H^2}{16\pi}\left(1-\frac{3\lambda}{a}\coth\frac{a}{\lambda}+\frac{3\lambda^2}{a^2}\right). \quad (6.96)$$

Consideration of the form of the variation of g with x shows that above a certain value of the radius a, supercooling and superheating should occur. Thus, if a is sufficiently large and H and T are within certain limits, the curve of g against x for given H and T has both a maximum and a minimum between $x=1$ and $x=0$, and the minimum cannot be reached from the side of $x=1$ unless the maximum is surmounted. The sphere therefore stays normal $(x=1)$

until H is reduced so much that the maximum just disappears. Similarly, if the metal is superconducting and H is being increased, the value $x = 1$ cannot be reached until the minimum at $x < 1$ has disappeared. It is easily shown that there is a critical size for supercooling and superheating to occur given by $(\partial g/\partial x)_{x=1} = 0$ and $(\partial^2 g/\partial x^2)_{x=1} = 0$; this yields

$$\frac{a}{\lambda_0} > \sqrt{\frac{21}{8}} \frac{T/T_c}{(1-(T/T_c)^2)^{\frac{1}{2}}}, \qquad (6.97)$$

as the condition for supercooling or superheating to occur. Although (6.97) shows that supercooling and superheating should occur for all sizes at $T = 0$, detailed calculations show that the supercooling and superheating would be very slight for spheres for which $a/\lambda_0 \ll 1$. Qualitatively, then, this argument explains why supercooling was found only for medium-sized spheres, but not for the smallest spheres in the colloid experiments (§ 5.2.1).

Calculation of the field for which $g = g_n$, which is equivalent for small enough spheres to finding the value of H for which

$$(\partial g/\partial x)_{x=1} = 0,$$

gives the critical field h of the sphere, and it is easily shown that for $a \ll \lambda_0$

$$\frac{h}{H_c} = \frac{\lambda_0}{a} \left(\frac{10}{1-(T/T_c)^2}\right)^{\frac{1}{2}}, \qquad (6.98)$$

in contrast to the result

$$\frac{h}{H_c} = \frac{\lambda_0}{a} \left(\frac{10}{1-(T/T_c)^4}\right)^{\frac{1}{2}}, \qquad (6.99)$$

obtained by applying (5.25) (in which λ is assumed independent of field). Lock (1950) has pointed out that (6.98) fits the colloid results (fig. 74 and § 5.8) much better than (6.99), if a is chosen so that either expression fits at low temperatures. The value of a has to be chosen as 0.6×10^{-6} cm., which is not perhaps unreasonable, since it is the critical field of the smallest particles which is measured, and the average size a' was estimated as 1.5×10^{-6} cm. (see § 5.2.1). Since, however, as already pointed out, neither of the corresponding relations for thin films can be fitted to the experimental data (a cannot of course be regarded as an adjustable parameter in the film experiments), it is dubious whether the improvement offered by (6.98) is of any great significance.

Further evidence for long-range order comes from the sharpness with which resistance disappears when a metal is cooled through the superconducting transition temperature. Pippard (1950c) assumes that during the temperature transition, domains of superconducting phase can be created by fluctuations only if their size exceeds a size characteristic of the long-range order. In this way the fraction of the material which is superconducting at temperatures close to T_c can be estimated, and hence (on the assumption that the super-conducting regions are randomly distributed spheres) the resistance is obtained as a function of temperature. From the sharpness of the sharpest observed transition (of order 10^{-3} ° K., by de Haas and Voogd (1931c)), Pippard deduces that the characteristic range is of order 10^{-4} cm., in agreement with the previous estimate.

The idea of long-range order also provides a clue to the signifi-cance of the surface energy discussed in Chapter IV. The boundary between normal and superconducting phases should not be regarded as a geometrical surface, but rather as a region in which the 'order' parameter (e.g. the x of the Gorter-Casimir theory) changes gradually. Owing to the long-range order inherent in the super-conducting state it is plausible to suppose that the width of this transition is of the same order of magnitude as the range of the long-range order. The magnetic field at the boundary will start to fall off at the point where the order parameter begins to rise, and will fall away to zero in a distance of order λ (but not equal to the penetration depth λ of a magnetic field at an interface between a superconductor and an insulator, since the change of order is more gradual at a normal-superconducting interface). These circum-stances are illustrated schematically in fig. 76; the effective 'boun-dary' between the phases can be set at A, where the order parameter is half-way between its extreme values. The line B represents the position such that if the field fell suddenly to zero at B the flux linkage would be equal to that which actually occurs (thus for exponential penetration, which would probably *not* apply here, B would be at distance λ from the beginning of penetration). If B were coincident with A, the total Gibbs free energy would be just that appropriate to the volumes to the left and right of A, on the assumption that the normal phase ceased abruptly at A and there was no penetration of field (i.e. g_n and $g_s + H_c^2/8\pi$ per unit volume respectively) and there would be no surface energy. If, however, as in fig. 76, B is to the left of A, there would be an *excess* of Gibbs free energy of amount $AB \times H_c^2/8\pi$ per unit area of the boundary, in addition to the volume contributions of g_n per unit volume up to A and $g_s + H_c^2/8\pi$ per unit volume beyond A.

The distance AB is thus just the characteristic length Δ of Chapter IV, and we may notice that, in agreement with experiment,

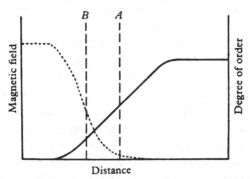

Fig. 76. Illustrating the physical meaning of the surface energy parameter Δ (Pippard, 1951 a). The broken curve is a schematic representation of the falling-off of the magnetic field crossing a boundary between normal and super-conducting phases; the full curve is a schematic representation of the increase of order.

Δ should be comparable to, though rather smaller than, 10^{-4} cm., the characteristic range of the long-range order. It is not surprising that Δ is much larger than β, the length defined in (5.23) as

$$\beta = 8\pi(\alpha_n - \alpha_s)/H_c^2,$$

where α_n and α_s are the surface energies of the two phases at an interface with an insulator; $\alpha_n - \alpha_s$ is a measure of how much the configuration of each phase is upset by a boundary with an insulator, and we have seen that it is very small. The length Δ, however, is associated primarily with the much larger disturbance near the interface of the two phases, owing to the impossibility of the order parameter changing too abruptly within the metal.

This interpretation of Δ suggests that if the range of the long-range order is much reduced—as in an alloy where the electronic mean free path is much shorter than in a pure metal—Δ might become negative, owing to B falling to the right instead of the left of A. As mentioned in §2.8 and explained more fully in §4.6.2

a negative value of Δ can go far to explain many of the peculiarities of superconducting alloys. Similarly, the local negative values of Δ required to explain the formation of superconducting nuclei in Faber's experiments might be expected to occur in regions where owing to inhomogeneous strain the range of order was small. Also it could be expected that Δ should be negative in very thin films, because the range of order would then be limited by the thickness of the film; this provides a possible mechanism for the formation of normal 'islands' which, as we saw in §5.8, is a possible cause for the rounding of the magnetization curves of thin films. Finally, it should be mentioned that the idea of persistence of order over a distance comparable with the film thickness provides the justification for ignoring in §5.8 the possibility of a boundary between normal and superconducting phases *parallel* to the film surface. In an earlier treatment by von Laue (1938, 1942) destruction of superconductivity in a film was supposed to proceed by such a boundary moving across the thickness of the film, and this leads to a smaller critical field than given by (5.25). For thicker films, of course, the formation of such a boundary is prevented by the existence of positive surface energy.

6.3.6. *Thermal conductivity.* The behaviour of the thermal conductivity of superconducting metals finds an immediate qualitative interpretation on the two-fluid model, though no detailed quantitative treatment is possible without special assumptions. According to the theory of metals, the electronic thermal conductivity κ is given by

$$\kappa \propto clv, \tag{6.100}$$

where c is the specific heat per unit volume of the electrons, l their mean free path, and v their velocity, provided the metal is in the residual resistance range where collisions are with irregularities rather than with thermal vibrations. When the metal becomes superconducting, only the 'normal' electrons can carry a heat current (since the entropy of the superconducting electrons is zero), and the thermal conductivity becomes modified in two ways by the condensation process which sets in as the transition temperature is passed. First, c is modified because there are fewer normal electrons, and secondly, the mean free path is effectively increased, because the number of free states to which normal electrons can go in a collision is reduced through the occupation of some of these states by condensed electrons. In calculating the change of c we must take no account of the contribution to the specific heat arising from the heat of evaporation of electrons from the superconducting to the normal phase, or in other words it is $T(\partial S/\partial T)_x$ we need rather than $T dS/dT$. This is because we are concerned in (6.100) merely with the transport of energy by electrons which remain 'normal' in passing along a temperature gradient, rather than with any change of their energy when they change their phase.* Thus

$$c = aTx^{\frac{1}{2}} = aT^3/T_c^2 \qquad (6.101)$$

(instead of $3aT^3/T_c^2$ which is the true specific heat including the extra heat due to transitions).

We should therefore expect to find

$$\frac{\kappa_s}{\kappa_n} = \left(\frac{T}{T_c}\right)^2 \frac{l_s}{l_n}, \qquad (6.102)$$

where l_s is the mean free path of the normal electrons in the super-conducting state and l_n the ordinary mean free path. No theory of the temperature variation of l_s can be given until the detailed theory of superconductivity is worked out, and it is in any case too much to hope that the rather naïve approach adopted here can do more than indicate qualitative trends. Heisenberg (1948 a,b) has suggested an interpolation formula for l_s of the form

$$l_s = l_n/\tfrac{1}{2}(1 + x). \qquad (6.103)$$

Clearly when $x = 1(T = T_c)$ this gives $l_s = l_n$, while for $x = 0(T = 0)$, $l_s = 2l_n$, corresponding in Heisenberg's theory to the fact that just half the states available to the few normal electrons left are blocked by 'superconducting' electrons. This gives

$$\frac{\kappa_s}{\kappa_n} = \frac{2(T/T_c)^2}{1 + (T/T_c)^4}, \qquad (6.104)$$

which is in fair agreement with the function f describing the experimental data (see §3.6). For metals with a low Debye Θ, scattering of electrons by lattice vibrations plays a dominant role, and the simple approach adopted here is no longer possible; no theory to account for the form of the empirical function $g(T)$ of

---

* It should be noticed that if c were the *actual* electronic specific heat there would be a discontinuity of κ at the superconducting transition. No trace of such a discontinuity was found by Hulm (1950) who made careful measurements in the transition region (see § 3.6).

§ 3.6 has yet been suggested. The two-fluid model can perhaps also explain very roughly why the function $h(T)$, which according to Hulm's interpretation represents the factor by which the lattice conductivity is multiplied when the metal becomes superconducting, is greater than unity. Thus, Hulm (1950) has suggested that since the dominant mechanism of thermal resistance in these conditions is that of scattering of lattice waves (phonons) by electrons, the mean free path for such scattering, and consequently the conductivity, will increase, when the number of electrons effective in scattering is reduced by the Gorter-Casimir condensation. Probably, however, this is too naïve an interpretation of the complicated mechanisms involved in lattice conduction, and, moreover, as we saw in § 3.6, it is doubtful whether it is possible to consider the thermal conductivity of superconducting alloys solely in terms of lattice conduction.

6.4. Fundamental theories

It has been supposed for a long time that the failure of the conventional electron theory of metals to explain superconductivity has been due to its neglect of the interaction between electrons. Thus the very complicated problem of an electron in a metal moving in the field of the other electrons and the positive ions is usually simplified to that of an electron moving in a periodic potential field. In this section we shall give a very brief descriptive account of some recent theories which take into account particular kinds of electronic interactions, and attempt to reproduce certain aspects of the known facts of superconductivity. The first of these theories is that of Heisenberg (1947, 1948 a, b; see also Koppe, 1950), who considers the ordinary Coulomb repulsion between electrons. He shows that if this interaction is introduced as a perturbation to the ordinary 'one-electron' solution of the problem, the resulting changes in the energy levels diverge; in other words, the effect of the interaction appears to be more serious than that of a small perturbation. Heisenberg then shows that below a certain temperature a state of slightly lower free energy is obtained if some of the electrons at the surface of the ordinary Fermi distribution in momentum space (the 'Fermi sphere') 'condense' into a lattice of wave-packets.

The energy of condensation according to a complicated but necessarily approximate calculation (Koppe, 1948) depends on a parameter Z, which is roughly the ratio of the kinetic energy of the electrons to the potential energy of two electrons an atomic distance apart. The dependence is, however, of an exponential kind, so that the energy of condensation turns out to be very much smaller than the Coulomb potential energy (which is of order $k \times 10^4 \,^\circ$ K.). When the temperature is raised above absolute zero, only a fraction ω of the Fermi sphere is covered by the condensed phase, and it turns out that this fraction ω plays the role of $(1 - x)$ of the Gorter-Casimir theory. The energy of condensation is thus related to the transition temperature in just the same way as in the Gorter-Casimir model, and the exponential dependence on Z can be adduced to explain (a) the rather wide spread in transition temperatures of the known superconductors—by a factor of order 30—in spite of the fact that Z would not be expected to vary nearly as much, and (b) the great sensitivity of transition temperature to strain (see p. 76).

According to this condensation picture, finite regions or domains of the metal have their momentum distribution unsymmetrical, or, in other words, the Fermi sphere has a 'bulge' on it corresponding to a permanent current.* The directions of the permanent current (i.e. the position of the bulge on the Fermi sphere) will, however, vary from domain to domain in such a way that there is no net current over any macroscopic region. It is only when a field is applied that the currents in the different domains line up to a certain extent to produce a macroscopic current. By consideration of this lining up process, both the Londons' equations are derived, but the arguments are rather *ad hoc*, and it is not very clear why the energy is not dissipated by the collisions of the electron lattices with the ordinary free electrons, since the argument is based on interactions between the two kinds of electrons which would seem to involve such dissipation.

* It should be noted that according to a theorem of Bloch and Peierls (unpublished but referred to by Brillouin (1933, 1935) and Bohm (1949)) the state of lowest free energy of a metal in the absence of applied fields is always one with no permanent current, and this may be raised as a fundamental objection to any theory of this kind. Heisenberg, however, argues that this theorem need not apply to microscopic regions, and is true only as an average over macroscopic regions.

An essential part of the theory is that at $0°$ K. the condensation into wave-packets takes place uniformly all round the Fermi sphere. There can therefore be no permanent current density at absolute zero; the detailed theory shows that as the temperature is lowered from T_c to zero, the permanent current density in an individual domain reaches a maximum value and then falls off to zero. This has the consequence that there should be a marked increase in the apparent penetration depth derived from the magnetization of a thin film when the magnetic field is sufficiently increased and the temperature sufficiently low; only in this way can the current density be kept low enough. As we have seen in §5.5, no such effect has been found experimentally, even though it has been looked for in conditions which, according to calculations by Koppe (1950), were particularly favourable. So far this prediction of increasing penetration depth for thin films at high fields and low temperature has been the only *new* prediction of the theory, and it is significant that it is at variance with the facts. Many criticisms have been levelled at the nature of the theory itself; an objection on fundamental grounds to the idea of permanent currents has already been mentioned (see footnote, p. 217), doubt has been cast on the validity of the approximations involved in the argument which shows that condensation into a lattice of wave-packets reduces the free energy, and the arguments which show that the condensed state has the properties of superconductivity have also been criticized. Perhaps the main value of the theory will prove to have been the revival, which it has undoubtedly stimulated, of active discussion of the problem.

A theory of somewhat similar type, but about which only little has been published, was proposed by Born and Cheng (1948). As the basis of this theory the Brillouin zone model of a metal was used rather than the simpler 'free electron gas' model used in Heisenberg's theory. Born and Cheng, again taking Coulomb interaction into account, found that the state of lowest free energy below a certain critical temperature is one in which some corners of the first Brillouin zone were filled and others were empty, leading again to an unsymmetrical distribution in momentum space. Here, however, the asymmetry of the distribution is still present at absolute zero, so that the difficulty of Heisenberg's theory, that it predicts no current at absolute zero, does not arise. One other feature of the Born and Cheng theory which makes it appear at first sight more attractive than Heisenberg's is that it is more specific in nature, for, by introducing the specific features of the Brillouin zone structure of individual metals, it is able to suggest why some metals are not superconductors. The Born and Cheng theory has, however, given no convincing explanation of how the proposed condensation into 'corners' can account for the actual properties of superconductors, and, just as in the Heisenberg theory, there is the objection on fundamental grounds to the idea of permanent currents, and the proof that a condensation actually occurs has not found universal acceptance.

Quite recently a new theory of superconductivity has been proposed by Fröhlich (1950, 1951a), and a similar theory, though somewhat differently developed, has also been proposed by Bardeen (1950, 1951a, b, c).* In these theories a quite different, and much smaller, interaction between electrons is considered; this is the indirect interaction brought about through the polarization of the ionic lattice by individual electrons. This polarization reacts back on the other electrons, and this causes a small interaction energy between electrons. It is rather questionable whether it is permissible to consider this smaller interaction while ignoring the much larger Coulomb interaction, but Fröhlich supposes that the Coulomb interaction is relevant only in discussing the mechanical cohesion of the metal and is unlikely to play any role in low-temperature phenomena. So far the effects of this interaction have been calculated only at absolute zero, and Fröhlich has shown that if a certain inequality is satisfied the interaction leads to removal of some of the electrons from the surface of the Fermi sphere to a spherical shell of slightly greater momentum. The energy gap, between these condensed electrons and the others, may be identified with the energy gap β in the two-fluid model, and in this way the transition temperature may be estimated. The values of T_c found are in reasonable order of magnitude agreement with the experimental values. As just mentioned, condensation takes place only if

---

* Mention should also be made of a theory by Tisza (1950, 1951), in which an attempt is made to treat the problem of electrons in a metal as a many-body problem; since however this theory has not been developed very far and does not lend itself easily to qualitative description, we shall not discuss it further.

a certain inequality is satisfied and so the theory predicts that only certain metals should be superconducting; moreover, apart from a few exceptions, it is in fact the right metals which are predicted. As is to be expected in view of the nature of the interaction mechanism involved, the formula for the transition temperature involves the mass of the positive ions, i.e. the atomic mass, M, of the metal. The detailed prediction is that T_c should vary as $M^{-\frac{1}{2}}$ if the other parameters, which involve only 'electronic' properties, are kept constant, and so this variation should be found as between the different isotopes of a given atomic species. As described in § 1.4.7, just this variation was in fact found experimentally, at about the same time as the theory was proposed.

The question of how far the Fröhlich theory reproduces the actual properties of superconductors has been only partly discussed so far. The main positive result is that by detailed consideration of how electrons behave in a magnetic field Fröhlich (1951 a) has been able to show that the condensation process is unable to occur in a magnetic field and hence that the state of lowest free energy is one in which the magnetic field is excluded. In fact, his theory leads to the Londons' equation (6.6), with an expression for λ which is of the same order as in (6.22), but involves also a dimensionless factor of order unity depending on the details of the interaction. Fröhlich's interpretation of the Meissner effect is in fact similar to the one proposed by London, the screening currents keeping the field out having the same character as the quantum currents inside atoms, but on a macroscopic scale (see § 6.2.2). Bardeen (1951 a) has also tried to explain the Meissner effect, but by the quite different approach of regarding it as a very large diamagnetism brought about by the electrons having very small effective mass. In both theories it is hoped that proof of the Meissner effect is a sufficient proof of superconductivity, though, as we have seen in § 6.2.2, it is not certain that the 'perfect conductivity' aspect can really be regarded as a subsidiary property. Neither theory has yet been developed to deal with finite temperatures.

TABLE VIII

The superconducting elements in the periodic system.

H																	He
Li 0·08 (7) cub.	Be 0·06 (1) hex.											B 1·3 (23) hex.	C 1·2 (22) hex.	N	O	F	Ne
Na 0·09 (7) cub.	Mg 0·05 (15) hex.											Al 1·20 (13,8) cub.	Si 0·07 (1) cub.	P 2·0 (15) orth.	S	Cl	A
K 0·08 (7) cub.	Ca 1·3 (23) cub.	Sc cub.	Ti 0·53 (20,4) hex.	V 5·1 (25,35) cub.	Cr 0·08 (1) cub.	Mn 0·15 (7) cub.	Fe 0·75 (14) cub.	Co 0·06 (26) hex.	Ni 0·75 (14) cub.	Cu 0·05 (15) cub.	Zn 0·91 (14,8) hex.	Ga 1·10 (10,8) orth.	Ge 0·05 (15) cub.	As 0·8 (12) rhom.	Se 1·3 (24) hex.	Br	Kr
Rb 0·8 (12) cub.	Sr 1·4 (23) cub.	Y 0·10 (7) cub.	Zr 0·70 (15) hex.	Nb 8 (21,11) cub.	Mo 0·05 (26) cub.	Tc	Ru 0·47 (7) hex.	Rh 0·09 (1) cub.	Pd 0·10 (7) cub.	Ag 0·05 (26) cub.	Cd 0·56 (15,8) hex.	In 3·37 (34,28) tetr.	Sn 3·730 (30,16) tetr.	Sb 0·15 (1) rhom.	Te 1·2 (24) hex.	I	Xe
Cs 0·8 (12) cub.	Ba 0·15 (7) cub.	La 4·37 (27,32) hex.	Hf 0·35 (15) hex.	Ta 4·4 (18,5) cub.	W 0·06 (26) cub.	Re 1·0 (2) hex.	Os 0·71 (7) hex.	Ir 0·10 (7) cub.	Pt 0·10 (7) cub.	Au 0·05 (15) cub.	Hg 4·152 (20,16) rhom.	Tl 2·38 (31,28) hex.	Pb 7·22 (30,3) cub.	Bi 0·05 (15) rhom.	Po mono.	At	Rn
Fr	Ra	Ac	Th 1·39 (19,33) cub.	Pa	U 0·8 (2,9) orth.												

Rare earths:

Ce 0·25 (7) cub.	Pr 0·25 (7) hex.	Nd 0·25 (7) hex.	Pm	Sm	Eu cub.	Gd hex.
Tb hex.	Dy hex.	Ho hex.	Er 0·8 (12) hex.	Tm hex.	Yb cub.	Lu hex.

Transuranic:

Np	Pu	Am	Cm

Notes to Table VIII

The elements enclosed in rectangles become superconducting at the temperatures indicated, and the others have been found not to become superconducting down to the temperatures indicated. The small figures are references to the literature; where two references are given the first refers to the original discovery and the second to the most recent or reliable value for T_c. The crystal system indicated is taken from Barrett (1943), where fuller details may be found. All temperatures are expressed in degrees absolute, based on the 1949 tables of vapour pressure of liquid helium (van Dijk and Shoenberg, 1949); the number of significant figures quoted is intended to indicate roughly the precision of the value given. The precision varies widely for a number of reasons which it is convenient to collect together here:

(1) T_c is sensitive to the quality of the specimen (see § 2.11), particularly for the hard superconductors, and in some cases quite different values have been obtained for different specimens. The following different estimates may be noted: Ti, $1\cdot21°$ K. (Meissner, 1930a), $1\cdot81°$ K. (Meissner, Franz and Westerhoff, 1932c) and $0\cdot53°$ K. (Daunt and Heer, 1949a; see also Shoenberg, 1940b); V, $4\cdot3°$ K. (Meissner and Westerhoff, 1934) and $5\cdot1°$ K. (Wexler and Corak, 1950b); Nb, between 7 and 9° K. (Jackson and Preston-Thomas, 1950; Cook, Zemansky and Boorse, 1950b); U, about $1\cdot3°$ K. (Aschermann and Justi, 1942; Alekseyevsky and Migunov, 1947), and about $0\cdot8°$ K. (Goodman and Shoenberg, 1950); Th, $1\cdot48°$ K. (Meissner, 1930b) and $1\cdot39°$ K. (Shoenberg, 1940b). The values quoted in the table are those corresponding to what is thought to be purest material.

(2) The transition is spread out to an extent depending on the quality of the specimen (see, for instance, fig. 3a, p. 5); T_c is usually defined as the temperature at which half the normal resistance is restored, and clearly this becomes the more uncertain the broader the transition.

(3) Precise determination of temperature involves a number of considerations which have not always been taken into account; we may mention the following: (a) in a long and narrow Dewar vessel there can be an appreciable pressure drop between the liquid level and the manometer on which pressure is recorded; this caused an appreciable error at low temperatures in some temperatures recorded by Shoenberg (1940b) (see Goodman and Mendoza, 1951); (b) the 1949 scale is expressed in terms of manometer mercury at 20° C., and a small correction (of order one- or two-thousandths of a degree) is necessary if the mercury temperature is much different from 20° C.; (c) a correction (which may amount to several thousandths of a degree) is necessary for the hydrostatic pressure of the liquid helium above the specimen, except below the λ-point (Lock, Pippard and Shoenberg, 1951; Atkins and Chase, 1951).

(4) The magnetic field of the measuring current in resistance measurements influences the transition (see fig. 3b, p. 5), and the presence of any other field (e.g. the earth's field) may both broaden and shift the transition.

So far these various factors have been adequately dealt with only for Sn and Hg, and it is only for these metals that the third decimal place in the transition temperature has any significance; even for these metals, discrepancies of nearly $0\cdot01°$ K. still remain between the results of different investigations (cf. various contributions on Sn and Hg isotopes at the Washington Low Temperature Symposium, 1951). Finally, it should not be forgotten that the conversion of helium vapour pressure to temperatures by means of the 1949 tables cannot be more reliable than the tables themselves, and departures of several thousandths of a degree from the true absolute temperatures due to systematic errors in the tables are not improbable.

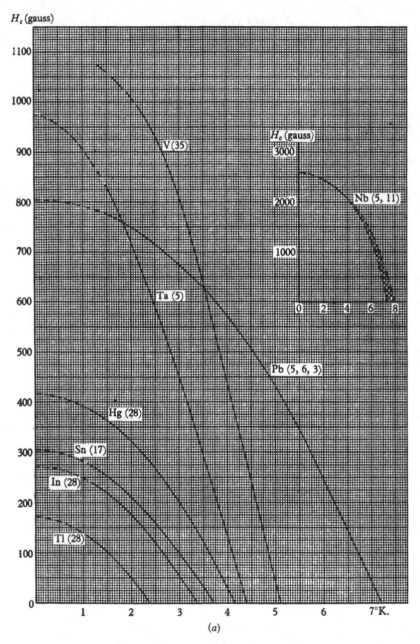

Fig. 77. H_c-T curves; for notes see p. 227.

H_c (gauss)

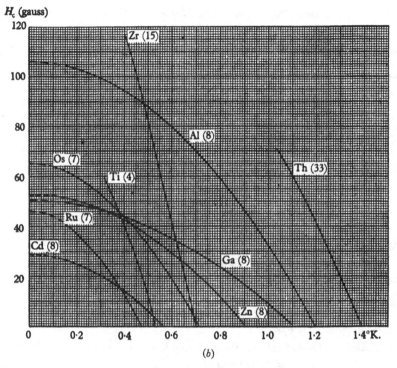

(b)

Fig. 77. H_c-T curves; for notes see p. 227.

References to Table VIII and figs. 77 a, b

(1) Alekseyevsky and Migunov (1947). (2) Aschermann and Justi (1942).
(3) Boorse, Cook and Zemansky (1950). (4) Daunt and Heer (1949a).
(5) Daunt and Mendelssohn (1937). (6) Daunt, Horseman and Mendelssohn (1939). (7) Goodman (1951a). (8) Goodman and Mendoza (1951).
(9) Goodman and Shoenberg (1950). (10) de Haas and Voogd (1929b).
(11) Jackson and Preston-Thomas (1950). (12) Justi (1946). (13) Keesom
(1933). (14) Keesom (1934). (15) Kürti and Simon (1935). (16) Laurmann and Shoenberg (1949). (17) Lock, Pippard and Shoenberg (1951).
(18) Meissner (1928). (19) Meissner (1929b). (20) Meissner (1930a).
(21) Meissner and Franz (1930a). (22) Meissner, Franz and Westerhoff
(1932c). (23) Meissner and Voigt (1930a). (24) Meissner and Voigt (1930b).
(25) Meissner and Westerhoff (1934). (26) Mendoza and Thomas (1951), and
private communication. (27) Mendelssohn and Daunt (1937). (28) Misener
(1940). (29) Onnes (1911b, c). (30) Onnes (1913c). (31) Onnes and Tuyn
(1922a). (32) Parkinson, Simon and Spedding (1951). (33) Shoenberg
(1940b). (34) Tuyn and Onnes (1923). (35) Wexler and Corak (1950b).

226SUPERCONDUCTIVITY

Table IX. *Numerical data for H_c*

For the critical fields of Hg, Sn, In and Tl the following empirical formulae have been given and have been used in constructing fig. 77; it should be noted, however, that the experimental data do not depart from the simple parabolic formula (1.1) by more than a few percent.

Hg (Misener, 1940):

$$H_c = 412 \cdot 58 - 19 \cdot 50 T^2 - 2 \cdot 133 T^3 + 0 \cdot 266 T^4.$$

This formula gives $T_c = 4 \cdot 167°$ K., which is slightly higher than the value given in Table VIII; the difference may be partly due to differences in specimen quality.

Sn (Lock, Pippard and Shoenberg, 1951):

$$H_c = 304 \cdot 5 (1 - 1 \cdot 0720(T/T_c)^2 - 0 \cdot 0944(T/T_c)^4$$
$$+ 0 \cdot 3325(T/T_c)^6 - 0 \cdot 1660(T/T_c)^8).$$

This formula may be used for any of the isotopes of tin if $304 \cdot 5$ is multiplied by $T_c/3 \cdot 730$, where T_c is the appropriate transition temperature.

In (Misener, 1940):

$$H_c = 269 \cdot 18 - 14 \cdot 86 T^2 - 6 \cdot 840 T^3 + 1 \cdot 299 T^4 - 0 \cdot 015 T^5.$$

This formula gives $T_c = 3 \cdot 368°$ K.

Tl (Misener, 1940):

$$H_c = 170 \cdot 67 - 30 \cdot 93 T^2 - 0 \cdot 193 T^3 + 0 \cdot 219 T^4.$$

This formula gives $T_c = 2 \cdot 380°$ K.

For Al, Ga, Zn, Os, Cd and Ru, the parabolic formula (see (1.1))

$$H_c = H_0(1 - (T/T_c)^2)$$

applies (Goodman, 1951*a*; Goodman and Mendoza, 1951).

For some of the other superconductors it is possible to estimate H_0, the value of H_c at $T = 0°$ K., by assuming the H_c-T curve is parabolic at sufficiently low temperatures, i.e. by linear extrapolation of a plot of H_c against T^2; the following values of H_0 are found (for Th, Ti and Zn the data do not go to low enough temperatures to permit extrapolation and only a rough extrapolation is possible for V); the values of T_c are given again here for convenience.

Element	Nb	Pb	V	Ta	Hg	Sn	In
T_c (° K.)	8	7·22	5·1	4·4	4·152	3·730	3·37
H_0 (gauss)	2600	805	∼1200	975	413	304·5	269

Element	Tl	Al	Ga	Zn	Os	Cd	Ru
T_c (° K.)	2·38	1·20	1·10	0·91	0·71	0·56	0·47
H_0 (gauss)	171	106	50·3	52·5	65	28·8	46

Of the remaining four superconducting elements, there are no data for Hf or Re; U has very high apparent values of H_c ($dH_c/dT \sim 2000$ gauss/° K.) (Goodman and Shoenberg, 1950), so too does La (~ 2000 gauss at $2 \cdot 3°$ K.) (Shoenberg, 1937*c*). As explained in the notes on p. 227 it is improbable that these are true equilibrium values.

Notes to figs. 77 a, b and Table IX

All the most reliable critical field data available at the time of writing have been included, and the reference numbers indicate the sources of the data used (see p. 225); the broken curves indicate extrapolations either by means of the formulae of Table IX or assuming the curve is parabolic at low temperatures. As far as possible all temperatures have been corrected to the 1949 scale (as in Table VIII), and most of the critical discussion in the notes to Table VIII applies here also. Critical fields may be determined either from resistance or magnetic measurements, but with alternating fields magnetic measurements may involve both resistive and magnetic properties. For 'ideal' superconductors, such as very pure monocrystalline tin, all methods give the same results (see, for instance, de Haas and Engelkes, 1937), but in specimens of poorer quality the results depend very much on the method as well as on the specimen; for Nb the range of variation for the specimen with the lowest critical fields is indicated by the shaded area (see also Cook, Zemansky and Boorse, 1950b). A valuable check on the reliability of critical field values as representing thermodynamic equilibrium fields, is a comparison with calorimetric data, and Table I (p. 62) lists all the superconductors for which a comparison with Rutgers's formula has been made; it can be seen that the high critical fields of La are clearly not equilibrium fields, but are of the same nature as the high critical fields of certain alloys (§ 2.8). It is probable that the high critical fields of U are also not equilibrium fields (this is suggested by rough calorimetric measurements by Goodman and Shoenberg (1950)), and it is possible that the critical fields of Ti, Zr, V and Nb are also somewhat higher than the true equilibrium fields.

It is convenient to mention here a number of measurements of critical fields not used in fig. 77 or Table IX, but of accuracy comparable to the data used, or of interest for other reasons: V (Webber, Reynolds and McGuire, 1949), Nb (Cook, Zemansky and Boorse, 1950b), Ta (Webber, 1947), Th (Shoenberg, 1938), Al (Keesom, 1933; Daunt and Heer, 1949b), Zn (Daunt and Heer, 1949b), In (Daunt, Horseman and Mendelssohn, 1939; Scott, 1948), Sn (de Haas and Engelkes, 1937; Daunt and Mendelssohn, 1937; Andrew, 1948a), Hg (Andrew, 1948a; Daunt and Mendelssohn, 1937), Tl (Daunt, Horseman and Mendelssohn, 1939).

TABLE X. *Superconducting compounds*

Compound	T_c (° K.)	Literature, remarks and values of dH_c/dT (gauss/° K.) at $T = T_c$
MoB ?	4·4	Hulm and Matthias (1951 *a*); queried by Ziegler and Young (1951)
NbB	6	Hulm and Matthias (1951 *a*)
ZrB	2·8–3·2	Meissner, Franz and Westerhoff (1932 *a*)
BaBi₃	6	Hulm and Matthias (1951 *b*)
Au₂Bi	1·7	de Haas, van Aubel and Voogd (1929 *a*); de Haas and Jurriaanse (1932); Jurriaanse (1935); de Haas and Voogd (1931 *b*); 100, Shoenberg (1938)
CaBi₃	1·7, 2·0	Alekseyevsky (1948 *b*); Alekseyevsky, Brandt and Kostina (1951)
KBi₂	3·6	160, Alekseyevsky (1950); 130, Reynolds and Lane (1950)
LiBi	2·5	Guttman and Stout (1951); Alekseyevsky, Brandt and Kostina (1951)
NaBi	2·2	100, Alekseyevsky (1949 *b*); 180, Reynolds and Lane (1950)
NiBi	4·2	Alekseyevsky, Brandt and Kostina (1951)
NiBi₃	3·6, 4·1	Alekseyevsky (1945 *c*); Alekseyevsky, Brandt and Kostina (1951)
PdBi₂	1·7	Alekseyevsky, Brandt and Kostina (1951)
RhBi	2·1	Alekseyevsky, Brandt and Kostina (1951)
RhBi₂	2·75, 2·9	Alekseyevsky (1948 *b*); Alekseyevsky, Brandt and Kostina (1951)
RhBi₄	2·2, 3·4	Alekseyevsky (1948 *b*); Alekseyevsky, Brandt and Kostina (1951)
SrBi₃	5·5	Hulm and Matthias (1951 *b*)
Mo-C	1·2–8·9	Meissner, Franz and Westerhoff (1932 *b*, 1933)
MoC	7·6–8·3	Meissner and Franz (1930 *b*); Meissner, Franz and Westerhoff (1933)
Mo₂C?	2·4–3·2	Meissner and Franz (1930 *b*); queried by Hudson (1951)
NbC	10·1–10·5	Meissner and Franz (1930 *b*)
TaC ?	9·3–9·5	Meissner and Franz (1930 *b*); queried by Ziegler and Young (1951)
TiC	1·1	Meissner and Franz (1930 *b*)
WC ?	2·5–4·2	Meissner and Franz (1930 *b*); queried by Ziegler and Young (1951)
W₂C	2·0–3·5	McLennan, Allen and Wilhelm (1931 *b*)
ZrC ?	2·3	Meissner, Franz and Westerhoff (1932 *a*); queried by Ziegler and Young (1951)
Nb-H	7–13	Aschermann, Friederich, Justi and Kramer (1941); Horn and Ziegler (1947)

TABLE X (cont.)

Compound	T_c (° K.)	Literature, remarks and values of dH_c/dT (gauss/° K.)
Ta-H	2–4	Horn and Ziegler (1947); Golik, Lasarew and Chotkewitsch (1949)
MoN	12	Hulm and Matthias (1951a)
Mo$_2$N	5	Hulm and Matthias (1951a)
NbN	15–16	~6000, Aschermann, Friederich, Justi and Kramer (1941); ~3300, Cook, Zemansky and Boorse (1950a); see also Andrews, Milton and De Sorbo (1946); Horn and Ziegler (1947)
TiN	1·2 and 5·5	Two jumps observed, Meissner, Franz and Westerhoff (1932a)
VN	1·5–3·2	Meissner and Franz (1930b)
ZrN	9·3– 9·6	Meissner and Franz (1930b); Meissner, Franz and Westerhoff (1932a)
CuS	1·6	Meissner (1929a); Buckel and Hilsch (1950); 130, Shoenberg (1938)
PbS?	4·2	McLennan, Allen and Wilhelm (1930b)
	5	Darby, Hatton and Rollin (1950); queried by Dunaev (1947); Hudson (1951); Hatton, Rollin and Seymour (1951)
PbSe	5	Darby, Hatton and Rollin (1950)
TaSi	4·4	Meissner, Franz and Westerhoff (1932a)
PbTe ?	5	Darby, Hatton and Rollin (1950); queried by Hudson (1951)

Notes to Table X

Included in this table are (a) compounds of two metals neither of which is superconducting, and (b) compounds of a metal with an insulator (borides, carbides, etc.); a few series of compounds of non-stoichometric proportions (Mo-C, Nb-H and Ta-H) are also listed here for convenience of reference, though they belong more properly in Table XI. A number of intermetallic compounds in which at least one component is a superconductor will be found listed in the table of alloy series (Table XI). A great many compounds have also been found non-superconducting to the lowest temperature tried; details will be found in the papers quoted in Table X. Many of the compounds listed have very broad transitions which vary very widely from one specimen to another, and in some cases rough limits are indicated for the specimens with the sharpest transitions. The contradictory results quoted for PbS and PbTe probably indicate that when superconductivity is observed it is due to free lead being present in the specimens.

TABLE XI. *Superconducting alloy series* (for notes see p. 232)

System	T_c (° K.)	Literature and remarks
Pb-Ag	5·8–7·3	de Haas, van Aubel and Voogd (1929b); McLennan, Allen and Wilhelm (1930b); Allen (1933)
Pb-As	8·4 (E)	McLennan, Allen and Wilhelm (1930b); queried by Meissner, Franz and Westerhoff (1933)
Pb-Au	2·0–7·3	McLennan, Allen and Wilhelm (1930b); Allen (1933); Pb₂Au 7·0, de Haas, van Aubel and Voogd (1929d)
Pb-Bi	7·3–8·8	McLennan, Allen and Wilhelm (1930a); Meissner, Franz and Westerhoff (1932d); high H_c for all compositions, particularly for E(8·8), (see fig. 14), de Haas and Voogd (1930, 1931b)
Pb-Ca	7·0	McLennan, Allen and Wilhelm (1930b)
Pb-Cd	7·0	McLennan, Niven and Wilhelm (1928); de Haas, van Aubel and Voogd (1929b); Schubnikow, Chotkewitsch, Schepelew and Rjabinin (1936); queried by Meissner, Franz and Westerhoff (1933)
Pb-Cu	7·3–7·8	Allen (1933)
Pb-Hg	4·1–7·3	For 15 % Hg, $H_c = 6800$ gauss at 4·2° K., de Haas and Voogd (1931b); Meissner, Franz and Westerhoff (1932b)
Pb-In	3·4–7·3	Meissner, Franz and Westerhoff (1932b); for 8 % In $H_c = 2400$ gauss at 4·2° K., Schubnikow, Chotkewitsch, Schepelew and Rjabinin (1936)
Pb-Li	7·2 (E)	McLennan, Allen and Wilhelm (1930b)
Pb₅Na₂	7·2	McLennan, Howlett and Wilhelm (1929)
Pb-P	7·8 (E)	McLennan, Allen and Wilhelm (1930b)
Pb-Sb	6·6 (E)	de Haas, van Aubel and Voogd (1929b); McLennan, Allen and Wilhelm (1930a)
Pb-Tl	2·2–7·3	Meissner, Franz and Westerhoff (1932d); for 40 % Tl and 30 % Tl $H_c \sim 3000$ gauss at 4·2° K., de Haas and Voogd (1931b), Schubnikow, Chotkewitsch, Schepelew and Rjabinin (1936)
PbTl₂	3·8	de Haas, van Aubel and Voogd (1930); high H_c (see fig. 16), de Haas and Voogd (1930); Rjabinin and Schubnikow (1935)
Sn-Ag	3·3–3·7	McLennan, Allen and Wilhelm (1932); Allen (1933)
Sn-As	4·1 (E)	McLennan, Allen and Wilhelm (1930b)
Sn-Au	2·6–3·7	Sn₂Au and Sn₄Au 2·5–2·75, de Haas, van Aubel and Voogd (1930); McLennan, Allen and Wilhelm (1932); Allen (1933)

TABLE XI (cont.)

System	T_c (° K.)	Literature and remarks
Sn-Bi	3·8 (E)	de Haas, van Aubel and Voogd (1929b); $H_c = 130$ gauss at 3·48° K., de Haas and Voogd (1929a)
Sn-Cd	3·6 (E)	de Haas, van Aubel and Voogd (1929b); $H_c = 266$ at 1·98° K., de Haas and Voogd (1929a)
Sn-Cu	3·6–3·7	Allen (1933)
Sn-Hg	4·2	Onnes (1913c) (this was the first superconducting alloy discovered)
SnSb	3·9	van Aubel, de Haas and Voogd (1928)
Sn_3Sb_2	4·0	van Aubel, de Haas and Voogd (1928); $H_c = 134$ gauss at 3·58° K., de Haas and Voogd (1930)
Sn-Tl	2·4–6·2	Meissner, Franz and Westerhoff (1932b); Allen (1933)
Sn-Zn	3·7	de Haas, van Aubel and Voogd (1929b); H_c approximately as for pure Sn, Lasarew and Nakhutin (1942)
Tl-Ag	2·7 (E)	de Haas, van Aubel and Voogd (1930); McLennan, Allen and Wilhelm (1932); queried by Meissner, Franz and Westerhoff (1933)
Tl-Au	1·9 (E)	de Haas, van Aubel and Voogd (1929b)
Tl_3Bi_5	6·4	van Aubel, de Haas and Voogd (1928); McLennan, Allen and Wilhelm (1930a); $H_c = 400$ gauss at 4·2° K. de Haas and Voogd (1929a)
Tl-Cd	2·5 (E)	de Haas, van Aubel and Voogd (1929b); queried by Meissner, Franz and Westerhoff (1933)
Tl_2Hg_5	3·8	de Haas, van Aubel and Voogd (1930)
Tl-In	2·4–3·7	Meissner, Franz and Westerhoff (1932b); magnetization curves with fair Meissner effect, Stout and Guttman (1950)
TlMg	2·75	$H_c = 220(1 - (T/2·75)^2)$, Guttman and Stout (1951)
Tl_7Sb_2	5·2	de Haas, van Aubel and Voogd (1929d); McLennan, Allen and Wilhelm (1930b)
Hg-Cd	1·7–4·1	McLennan, Niven and Wilhelm (1928); de Haas and de Boer (1932); for up to 10 % Cd, H_c approximately the same as for pure Hg, Schubnikow, Chotkewitsch, Schepelew and Rjabinin (1936)

Some superconducting alloys with more than two components are: Rose's metal 8·5, Newton's metal 8·5 and Wood's metal 8·2 (McLennan, Allen and Wilhelm, 1930a), PbAsBi, 9·0, PbAsBiSb 9·0 (McLennan, Allen and Wilhelm, 1930b).

Notes to Table XI

The temperature range given is usually that of the temperature at which half the resistance disappears (in a few cases, that at which the whole resistance disappears), for different alloys of the given series. Since the resistance disappears over a temperature interval sometimes as large as $2°$ K., which depends on the strength of the measuring current and the heat treatment of the alloy (details of these are not always to be found in the original papers), the figures given are necessarily rough. The letter E after a figure indicates that this is the mean transition temperature of the eutectic; apart from this, the table gives no indication of the type or structure of the various alloys; such information can sometimes be found in the original papers, which should be consulted in conjunction with the equilibrium diagrams of the alloy systems (see for instance Hansen, 1936). In some cases, it is likely that superconductivity is due to one of the components in the free state (e.g. Pb in the Pb-Cd series)—there are probably many more such cases than those indicated as 'queried'. The H_c values given are the field strengths at which half the resistance (sometimes the whole resistance) is restored; for a few alloys they are deduced from magnetization curves. These values are meant to serve only as a guide to the orders of magnitude; more detailed data can in some cases be found in the original papers. The original papers list also many particular alloys which have not been found superconducting. Although the table is fairly comprehensive it is not exhaustive; incidental data on alloys may be found also in papers primarily concerned with specific heats and thermal conductivity. For a review of most of the Leiden experiments on superconducting alloys see de Haas and Voogd (1932).

APPENDIX II

Some special solutions of the Londons' equations (see § 6.2.1)

(a) *Plate of thickness 2a in a uniform field H_0 parallel to its surface*

At distance z from the median plane the field is

$$H = H_0 \frac{\cosh z/\lambda}{\cosh a/\lambda}. \tag{A 1}$$

The corresponding current distribution follows by differentiation $\left(\text{since here } J = \frac{c}{4\pi} \frac{\partial H}{\partial z} \right)$. The magnetic moment I per unit volume of the plate is most simply obtained from

$$I = \frac{1}{8\pi a} \int_{-a}^{a} (H - H_0)\, dz,$$

which gives

$$I = -\frac{H_0}{4\pi} \left(1 - \frac{\lambda}{a} \tanh \frac{a}{\lambda} \right), \tag{A 2}$$

or, if we put $\kappa = I/H_0$ and $\kappa_0 = -1/4\pi$,

$$\frac{\kappa}{\kappa_0} = 1 - \frac{\lambda}{a} \tanh \frac{a}{\lambda}. \tag{A 3}$$

As has already been mentioned (see § 5.2) this simplifies to

$$\kappa/\kappa_0 = 1 - \lambda/a \quad \text{for } a \gg \lambda \tag{A 4}$$

and

$$\kappa/\kappa_0 = \tfrac{1}{3} a^2/\lambda^2 \quad \text{for } a \ll \lambda. \tag{A 5}$$

(b) *Cylinder of radius a in a uniform field H_0 parallel to its axis*

At distance r from the axis, the field is

$$H = H_0 J_0(jr/\lambda)/J_0(ja/\lambda), \tag{A 6}$$

where J_0 denotes the Bessel function of order zero and j is $\sqrt{-1}$. The magnetization is given by

$$I = \frac{1}{2\pi a^2} \int_0^{} (H - H_0)\, r\, dr,$$

which gives

$$I = -\frac{H_0}{4\pi} \left(1 + 2j \cdot \frac{\lambda}{a} \frac{J_1(ja/\lambda)}{J_0(ja/\lambda)} \right),$$

and
$$\kappa/\kappa_0 = 1 + 2j\,\frac{\lambda}{a}\,\frac{J_1(ja/\lambda)}{J_0(ja/\lambda)}. \tag{A 7}$$

This reduces to
$$\kappa/\kappa_0 = 1 - 2\lambda/a \quad \text{for } a \gg \lambda \tag{A 8}$$

and
$$\kappa/\kappa_0 = \tfrac{1}{8}a^2/\lambda^2 \quad \text{for } a \ll \lambda. \tag{A 9}$$

(c) Sphere of radius a in a uniform field H_0

Using spherical polar co-ordinates with the field direction as the axis of reference, the solution is:

For $r \leqslant a$:

$$H_r = 3H_0\,\frac{\lambda a}{r^2}\left(\frac{\sinh r/\lambda}{\sinh a/\lambda}\right)\left(\coth\frac{r}{\lambda}-\frac{\lambda}{r}\right)\cos\theta,$$

$$H_\theta = -\frac{3H_0}{2}\,\frac{\lambda a}{r^2}\left(\frac{\sinh r/\lambda}{\sinh a/\lambda}\right)\left(\coth\frac{r}{\lambda}-\frac{\lambda}{r}\left(1+\frac{r^2}{\lambda^2}\right)\right)\sin\theta.$$

$$H_\phi = 0. \tag{A 10}$$

For $r \geqslant a$:
$$H_r = (H_0 + 2M/r^3)\cos\theta,$$
$$H_\theta = (-H_0 + M/r^3)\sin\theta,$$
$$H_\phi = 0, \tag{A 11}$$

where
$$M = -\frac{H_0 a^3}{2}\left(1 - \frac{3\lambda}{a}\coth\frac{a}{\lambda} + \frac{3\lambda^2}{a^2}\right). \tag{A 12}$$

Equation (A 12) means that the field outside the sphere is the original one together with that of a dipole M. Thus M is just the induced magnetic moment of the sphere, and we obtain for κ/κ_0 (where $\kappa_0 = -3/8\pi$ in accordance with equation (2.16)):

$$\frac{\kappa}{\kappa_0} = 1 - \frac{3\lambda}{a}\coth\frac{a}{\lambda} + \frac{3\lambda^2}{a^2}. \tag{A 13}$$

This reduces to
$$\kappa/\kappa_0 = 1 - 3\lambda/a \quad \text{for } a \gg \lambda, \tag{A 14}$$

and
$$\kappa/\kappa_0 = \tfrac{1}{15}a^2/\lambda^2 \quad \text{for } a \ll \lambda. \tag{A 15}$$

(d) Infinite cylinder of circular cross-section (radius a) carrying total current i

The current density at radius r is given by

$$J = \frac{i}{2\pi a\lambda}\,j\,\frac{J_0(jr/\lambda)}{J_1(ja/\lambda)}, \tag{A 16}$$

where J_0 and J_1 are Bessel functions and $j = \sqrt{-1}$.

For $a \gg \lambda$, (A 16) reduces to

$$J = \frac{i}{2\pi a \lambda}\, e^{-(a-r)/\lambda}. \qquad\qquad \text{(A 17)}$$

So that the maximum current density is

$$J_{\max.} = i/2\pi a \lambda. \qquad\qquad \text{(A 18)}$$

For $a \ll \lambda$, (A 16) reduces to a uniform distribution

$$J = i/\pi a^2. \qquad\qquad \text{(A 19)}$$

APPENDIX III

Notes on recent work (added 1 May 1952)

CHAPTER I

p. 12. An isotope effect of the expected order of magnitude has been found in lead (Olsen-Bär, 1951; Reynolds and Serin, 1951), and thallium (Maxwell, 1951).

CHAPTER III

p. 82. Goodman (1951 b) has extended the measurements of κ_{es}/κ_{en} for tin down to 0.2° K. and finds that $f(T/T_c)$ agrees well with a more exact version of Heisenberg's formula (6.104) (see p. 215) based on a calculation by Koppe (1947). According to this calculation x falls more rapidly than $(T/T_c)^4$ at low temperatures, and this makes both κ_{es} and c fall off more rapidly with temperature than predicted by (6.104) and (6.47) respectively.

p. 85. Webber and Spohr (1951) have found effects in pure lead similar to those shown in fig. 31 a for impure lead; the minimum of κ in the intermediate state is thus probably not due to impurities.

p. 93. Pullan (1951), using a sensitive superconducting galvanometer (Pippard and Pullan, 1952), has shown that the thermoelectric power of tin does behave very nearly in the 'ideal' manner indicated in fig. 35. For a pure single-crystal specimen with its tetragonal axis parallel to the temperature gradient he finds also that the value of α (see (3.36) and Table III) is 0.22×10^{-8} V./(° K.)², i.e. much smaller than the value found by Steele (1951) and Webber and Steele (1950), but agreeing fairly well with the value of Casimir and Rademakers (1947). Pullan's experiments, which promise to clear up the confused situation described in the text, are being continued with a view to investigating how α depends on purity and crystal orientations.

CHAPTER IV

p. 100. Lifshitz and Sharvin (1951) have shown that it is only for fairly large specimens that the free energy of Landau's branching model becomes lower than that of his earlier non-branching model. They find that for specimens of order 2 cm. thick each normal lamina should branch only once or twice, and that these branches should come out to the surface in the manner indicated in fig. 36. They have also corrected what appears to be a numerical slip in one of the equations given by Landau (1937); the corrected version is quoted in (4.4).

p. 118. Chambers (unpublished) has pointed out an error in the thermodynamic arguments used by Kuper (1951). This error is likely to have little effect on Kuper's estimates of Δ based on the ρ values but may be responsible for the very high values of Δ estimated from the slopes of the falling parts of the magnetization curves.

p. 129. Garfunkel and Serin (1952) have made experiments in which the central region of a long superconducting tin rod was in a higher magnetic field than the ends. They found in this way that for temperatures close to T_c the central region could be considerably 'superheated' (to above 20% above H_c). Their results suggest that the difficulty of observing 'superheating' in Faber's experiments is associated with the formation of normal nuclei at the ends of the rod.

CHAPTER V

p. 161. A brief note by Sharvin (1951) reports a considerable increase of penetration depth with magnetic field (λ varying as $(1 + 0.28 (H/H_c)^2)$. This is a much larger effect than found by Pippard (1950c), and, since unsuitable specimen conditions can easily give an exaggerated estimate of the field dependence, Sharvin's result must be regarded with reserve until fuller details of his experiments are available.

p. 167. Shalnikov's experiments (1940a) on thin tin films were of an exploratory character and only a few quantitative data on h/H_c were obtained for annealed films; Alekseyevsky's results (fig. 71 b) too were not very complete. Fuller data for thin tin films have recently been obtained by Zavaritski (1951) using Shalnikov's technique. For films thicker than 3×10^{-4} cm. (5.18) was found to hold but with $b = 7 \pm 1 \times 10^{-6}$ cm., i.e. appreciably lower than Lock's and Andrew's values (see p. 168). For the thinner films the values of h/H_c agree roughly with those of Alekseyevsky (see

238 SUPERCONDUCTIVITY

fig. 71 b), though they tend to be about 30% lower; the general character of the variation of h/H_c with thickness and temperature is similar to that found by Appleyard *et al.* for mercury (see fig. 71 a).

CHAPTER VI

p. 204. Pippard (to be published) has extended his radio-frequency measurements on tin contaminated with indium to specimens of still greater indium content, and found that the penetration depth λ increases appreciably for the largest indium contents; corresponding changes in R/R_n as a function of temperature were also observed. Pippard interprets the increase of λ with impurity content as indicating the need for modifying the Londons' basic equation (6.6) to a more complicated integral form involving the range of long-range order (see p. 207); in terms of this interpretation the London theory should apply only as a limiting case for very impure specimens, in which the range has become shorter than λ.

p. 207. Ginsburg and Landau (1950) have developed a modification of the Londons' phenomenological theory in order to account for positive surface energy in a more fundamental way and to deal more adequately with problems such as the destruction of superconductivity in thin films by magnetic fields. Their theory leads to the concept of long-range order, and some of its consequences are therefore similar to those arising from Pippard's ideas discussed in § 6.3.5. A function ψ is introduced such that $|\psi|^2$ may be considered as the density of 'superconducting' electrons. The free energy in a magnetic field is assumed to contain a term proportional to $|\operatorname{grad}\psi|^2$ with the consequence that energy contributions arise not only from the kinetic energy of the super-current but also from any spatial variation of the concentration of superconducting electrons; this produces a tendency to long-range order so that, for instance, at a boundary between normal and superconducting phases there is a relatively gradual transition of $|\psi|^2$. The form of this transition is calculated by minimization of the free energy, and hence the interphase surface energy is evaluated in terms of the critical field and the penetration depth, but no adjustable parameters. The calculations apply only close to T_c where reasonable agreement with experiment is obtained. The field variation of penetration depth is then calculated and found to be very small in agreement with Pippard's experimental results. The variation of h/H_c for thin films with thickness and temperature is worked out and approximate agreement with the results of Appleyard *et al.* is found (though a rather liberal interpretation of the data for λ is required). Zavaritski (see above) also interprets his experiments on tin in terms of this theory and finds fair agreement. Some discussion of the destruction of superconductivity of thin films by a current is also given, but the edge effects mentioned on p. 177 appear to be ignored. Some aspects of the theory are discussed in more detail by Ginsburg (1950), who has also (1951) discussed the behaviour of superconductors in radio-frequency fields in terms of the theory.

p. 219. A number of papers have appeared recently discussing various aspects of Fröhlich's and Bardeen's theories. Wentzel (1951) has criticized the perturbation method used in these theories and suggests that the interaction assumed would make the crystal lattice unstable (see also Kohn and Vachaspati, 1951; Drell (1951), and Klein (1952)), but his objections have been answered by Huang (1951). The explanation of the Meissner effect offered by Fröhlich has been criticized by Schafroth (1951), while Fröhlich (1951 *b*) has criticized Bardeen's explanation. We may mention also a recent review article by Bardeen (1951 *d*) in which some of the above criticisms are discussed. In view of this discussion it appears that the account of the theories given in the text, and the hopes expressed in the Preface, are perhaps overoptimistic.

APPENDIX I

p. 222. Hilsch (1951) has found that bismuth films if evaporated on to quartz at $4 \cdot 2°$ K. become superconducting with $T_c \sim 6°$ K. After annealing at high temperatures superconductivity is no longer observed. It is probable that the superconductivity is associated with the amorphous structure of the unannealed films. This result confirms the idea that ordinary bismuth is in some way nearly a superconductor, but that superconductivity actually occurs only if the bismuth is modified either by compounding with another element (see Table X) or by a change of its atomic arrangement, as in Hilsch's films. Alekseyevsky (1949 *a*) has suggested that ordinary bismuth might become superconducting at sufficiently high pressures.

Note on developments since 1952

In the preface it was suggested that superconductivity might not 'much longer remain in its hitherto impregnable position of an unsolved mystery of science'. This optimism has been justified by the recent theoretical work of Bardeen, Cooper and Schrieffer and of Bogolyubov, Valatin and others, which has at long last provided a fundamental basis for understanding the observed facts. The interaction between electrons responsible for 'condensation' into the superconducting state proves to be an interaction through the phonons as in Fröhlich's and Bardeen's earlier theories (mentioned briefly on p. 219), but with the difference that it is interaction between electrons of opposite momenta and spins which is of major importance. An interesting feature of the new theory is that it provides a natural explanation of the non-local electrodynamics originally proposed by Pippard on empirical grounds (p. 238).

On the experimental side there have been some important advances also. Experiments on the behaviour of superconductors in high frequency alternating fields have been extended into the region of much higher frequencies and have provided direct evidence for the existence of an energy gap in the excitation spectrum of superconductors. The presence of the gap, which is required by the theory, had earlier been indicated by precision measurements of the electronic specific heat in the superconducting state, which showed an exponential variation at very low temperature. Studies of nuclear resonance, nuclear spin relaxation, and of the absorption of ultrasonic waves in the superconducting state have provided interesting new evidence, most of which is explained at least qualitatively by the new theory, though some of it suggests that considerable refinement of the theory is still required. There has also been a considerable accumulation of new quantitative data relating to phenomena already discussed in the main body of the text. An interesting development has been the invention of superconducting computer elements which promise to be of technical importance; this has stimulated a considerable amount of work on thin superconducting films and indeed on superconductivity in general.

To help in following the developments since 1952 a brief bibliography of important review articles and conference publications is appended:

Theory

BARDEEN, J. (1956). *Encyclopedia of Physics*, **15**, 274. Berlin: Springer. (Review of progress up to 1956.)

BARDEEN, J., COOPER, L. N. and SCHRIEFFER, J. R. (1957). *Phys. Rev.* **108**, 1175. (This is the paper in which the definitive theory was first presented.)

BOGOLIUBOV, N. N., TOLMACHEV, V. V. and SHIRKOV, D. V. (1958). *A new method in the theory of superconductivity.* Moscow: Academy of Sciences of the U.S.S.R. (in Russian). English Translation in *Fortschr. Phys.* **6**, 605 (1958).

KHALATNIKOV, I. M. and ABRIKOSOV, A. A. (1959). The modern theory of superconductivity. *Phil. Mag. Suppl.* **8**, 45. (Review in which applications of the theory are worked out using Bogoliubov's methods.)

KUPER, C. G. (1959). The theory of superconductivity. *Phil. Mag. Suppl.* **8**, 1. (Comprehensive general review.)

Experimental

BIONDI, M. A., FORRESTER, A .T., GARFUNKEL, M. P. and SATTERTHWAITE, C. B. (1958). Experimental Evidence for an energy gap in superconductors. *Rev. Mod. Phys.* **30**, 1109. (Review of thermal properties, nuclear effects, ultrasonic absorption and infra-red and microwave results.)

PIPPARD, A. B. (1954). *Advances in Electronics and Electron Physics*, **6**, 1. New York: Academic Press. (Review of microwave results up to 1954.)

RHODERICK, E. H. (1959). Superconducting computer elements. *Br. Jour. App. Phys.* **10**, 193.

SERIN, B. (1956). *Encyclopedia of Physics*, **15**, 210. Berlin: Springer. (General review.)

Conferences

Proceedings of the 5th International Conference of Low Temperature Physics and Chemistry, Madison, Wisconsin 1957. University of Wisconsin Press, 1958.

Proceedings of the Kamerlingh Onnes Conference on Low Temperature Physics, Leiden 1958. Supplement to *Physica*, **24**, 1958.

BIBLIOGRAPHY AND AUTHOR INDEX

Except for the list of earlier general articles and books on superconductivity, this bibliography (which serves also as an author index) lists
only those papers which are explicitly referred to in the text. Usually the
papers chosen for reference are the first to discuss the point in question
and the most recent; intermediate papers are mentioned only when they
present some special feature of interest. Preliminary notes, such as letters
to *Nature*, are quoted only where the date of a discovery is involved or
when the full paper is not easily accessible (as with papers in the Russian
language). More comprehensive bibliographies may be found in some of
the general works listed below (Steiner and Grassmann, 1937; Ginsburg,
1946; and Justi, 1948). References to the early Leiden papers are
given by the number of the Leiden Communication (usually in English),
since the original papers were usually in Dutch; since 1934, however, first
publication has been usually in English, in *Physica*. Until 1946, most
Russian papers appeared both in the Russian language and translated into
English or German, and only the reference to the translated version is
given; since then, however, no translated versions have been published,
and the reference is to the Russian language periodical. To facilitate
reference, the spelling of Russian names follows the author's own transliteration, where any papers by him have appeared in an English or
German language periodical, even though this transliteration sometimes
follows German rather than English rules; where the author's publications
have appeared only in Russian, English rules of transliteration have been
used.

(1) *Some general references*

(a) *Articles and books*

BURTON, E. F. and others. *The Phenomenon of Superconductivity*. Toronto,
1934.
BURTON, E. F., SMITH, H. GRAYSON and WILHELM, J. O. *Phenomena at
the Temperature of Liquid Helium*, Chapters 5 and 10. New York,
1940.
GINSBURG, V. L. *Superconductivity* (in Russian). Moscow, Leningrad, 1946.
GINSBURG, V. L. *Usp. Fiz. Nauk*, **12**, 169 and 133, 1950.
DE HAAS, W. J. *Leipziger Vorträge*, 1933, p. 59.
JACKSON, L. C. *Low Temperature Physics*, Chap. v, 2nd ed. London,
1948.
JUSTI, E. *Leitfähigkeit und Leitungsmechanismus fester Stoffe*, Chap. VII.
Göttingen, 1948.
VON LAUE, M. *Theorie der Supraleitung*, 2nd ed. Berlin, Gottingen,
Heidelberg, 1949.
LONDON, F. *Une conception nouvelle de la supraconductibilité*. Paris, 1937.
LONDON, F. *Superfluids*, vol. 1, New York, 1950.
MEISSNER, W. *Ergebn. Exakt. Naturw.* 1932, **11**, 219.
MEISSNER, W. *Handb. Exp. Phys.*, 1935, **11** (pt. 2), 204.

MENDELSSOHN, K. *Rep. Progr. Phys.* 1949, **12**, 270.
ONNES, K. H. *Commun. Phys. Lab. Univ. Leiden*, Suppl. no. 50a, 1924 and no. 66a, 1928.
SMITH, H. GRAYSON and WILHELM, J. O. *Rev. Mod. Phys.* 1935, **7**, 238.
STEINER, K. and GRASSMANN, P. *Supraleitung.* Braunschweig, 1937.

(b) *Some Conferences (since 1935)*
1935 *Proc. Roy. Soc.* A, **152**, 1.
1937 *Actes du VIIᵉ Congrès International du Froid.*
1946 *International Conference on Fundamental Particles and Low Temperatures*, Cambridge, vol. II.
1949 *International Conference on Low Temperature Physics*, M.I.T.
1951 *Low Temperature Symposium*, National Bureau of Standards, Washington.
1951 *Oxford Conference on Low Temperatures.*

(2) *Original papers and books quoted in the text*
The figures in italics indicate the pages of the text in which the reference is quoted.

ALEKSEYEVSKY, N. E. (1938a). *J. Exp. Theor. Phys. U.S.S.R.* (Russian), **8**, 342. [*134*]
ALEKSEYEVSKY, N. E. (1938b). *J. Exp. Theor. Phys. U.S.S.R.* (Russian), **8**, 1098. [*43*]
ALEKSEYEVSKY, N. E. (1940). *J. Phys. U.S.S.R.* **3**, 443. [*75*]
ALEKSEYEVSKY, N. E. (1941a). *C.R. Acad. Sci. U.R.S.S.* **32**, 31. [*36*]
ALEKSEYEVSKY, N. E. (1941b). *J. Phys. U.S.S.R.* **4**, 401. [*54, 167, 169*]
ALEKSEYEVSKY, N. E. (1945a). *J. Phys. U.S.S.R.* **9**, 147. [*75*]
ALEKSEYEVSKY, N. E. (1945b). *J. Phys. U.S.S.R.* **9**, 217. [*128*]
ALEKSEYEVSKY, N. E. (1945c). *J. Phys. U.S.S.R.* **9**, 350. [*38, 228*]
ALEKSEYEVSKY, N. E. (1946). *J. Phys. U.S.S.R.* **10**, 360. [*36, 120*]
ALEKSEYEVSKY, N. E. (1948a). *Doklady Akad. Nauk S.S.S.R.* (Russian), **60**, 37. [*128*]
ALEKSEYEVSKY, N. E. (1948b). *J. Exp. Theor. Phys. U.S.S.R.* (Russian), **18**, 101. [*38, 228*]
ALEKSEYEVSKY, N. E. (1949a). *J. Exp. Theor. Phys. U.S.S.R.* (Russian), **19**, 358. [*76, 239*]
ALEKSEYEVSKY, N. E. (1949b). *J. Exp. Theor. Phys. U.S.S.R.* (Russian), **19**, 671. [*38, 228*]
ALEKSEYEVSKY, N. E. (1950). *J. Exp. Theor. Phys. U.S.S.R.* (Russian), **20**, 863. [*38, 228*]
ALEKSEYEVSKY, N. E., BRANDT, N. B. and KOSTINA, T. I. (1951). *J. Exp. Theor. Phys. U.S.S.R.* (Russian), **21**, 951. [*228*]
ALEKSEYEVSKY, N. E. and MIGUNOV, L. (1947). *J. Phys. U.S.S.R.* **11**, 95. [*223, 225*]
ALLEN, J. F. (1933). *Phil. Mag.* **16**, 1005. [*230, 231*]
ALLEN, W. D., DAWTON, R. H., BÄR, M., MENDELSSOHN, K. and OLSEN, J. L. (1950). *Nature, Lond.*, **166**, 1071. [*12*]

ALLEN, W. D., DAWTON, R. H.., LOCK, J. M., PIPPARD, A. B. and
  SHOENBERG, D. (1950). *Nature, Lond.*, **166**, 1071. [*12*]
ANDREW, E. R. (1948*a*). *Proc. Roy. Soc.* A, **194**, 80. [*112, 113, 116, 118,*
  227]
ANDREW, E. R. (1948*b*). *Proc. Roy. Soc.* A, **194**, 98. [*98, 101, 103, 114, 115*]
ANDREW, E. R. (1949). *Proc. Phys. Soc.* A, **62**, 88. [*168*]
ANDREW, E. R. and LOCK, J. M. (1950). *Proc. Phys. Soc.* A, **63**, 13.
  [*36, 114, 117, 118, 119, 120*]
ANDREWS, D. H., MILTON, R. M. and DE SORBO, W. (1946). *J. Opt. Soc.*
  Amer. **36**, 518. [*229*]
APPLEYARD, E. T. S. and BRISTOW, J. R. (1939). *Proc. Roy. Soc.* A, **172**,
  530. [*148*]
APPLEYARD, E. T. S., BRISTOW, J. R. and LONDON, H. (1939). *Nature,*
  Lond., **143**, 433. [*166*]
APPLEYARD, E. T. S., BRISTOW, J. R., LONDON, H. and MISENER, A. D.
  (1939). *Proc. Roy. Soc.* A, **172**, 540. [*138, 166, 168*]
APPLEYARD, E. T. S. and MISENER, A. D. (1938). *Nature, Lond.*, **142**, 474.
  [*138, 166*]
ARKADIEV, V. (1945). *J. Phys. U.S.S.R.* **9**, 148. [*19*]
ARKADIEV, V. (1947). *Nature, Lond.*, **160**, 330. [*19*]
ASCHERMANN, G., FRIEDERICH, E., JUSTI, E. and KRAMER, J. (1941).
  Phys. Z. **42**, 349. [*228, 229*]
ASCHERMANN, G. and JUSTI, E. (1942). *Phys. Z.* **13**, 207. [*223, 225*]
ATKINS, K. R. and CHASE, C. E. (1951). *Proc. Phys. Soc.* A, **64**, 826. [*223*]
VAN AUBEL, E., DE HAAS, W. J. and VOOGD, J. (1928). *Commun. Phys.*
  Lab. Univ. Leiden, no. 193*c*. [*231*]
BARDEEN, J. (1950). *Phys. Rev.* **80**, 567. [*180, 185, 219*]
BARDEEN, J. (1951*a*). *Phys. Rev.* **81**, 829. [*219, 220*]
BARDEEN, J. (1951*b*). *Phys. Rev.* **81**, 1070. [*219*]
BARDEEN, J. (1951*c*). *Phys. Rev.* **82**, 978. [*219*]
BARDEEN, J. (1951*d*). *Rev. Mod. Phys.* **23**, 261. [*239*]
BARRETT, C. S. (1943). *Structure of Metals.* New York and London:
  McGraw Hill. [*223*]
BECKER, R., HELLER, G. and SAUTER, F. (1933). *Z. Phys.* **85**, 772. [*180,*
  192]
BOHM, D. (1949). *Phys. Rev.* **75**, 502. [*217*]
BOORSE, H. A., COOK, D. B. and ZEMANSKY, M. W. (1950). *Phys. Rev.*
  78, 635. [*225*]
BORELIUS, G., KEESOM, W. H., JOHANSSON, C. H. and LINDE, J. O. (1931).
  Commun. Phys. Lab. Univ. Leiden, no. 217*c*. [*86*]
BORELIUS, G., KEESOM, W. H., JOHANSSON, C. H. and LINDE, J. O. (1932).
  Commun. Phys. Lab. Univ. Leiden, Suppl. no. 70*a*. [*88*]
BORN, M. and CHENG, K. C. (1948). *Nature, Lond.*, **161**, 968 and 1017;
  J. Phys. Radium, **9**, 249. [*180, 218*]
BREMMER, H. and DE HAAS, W. J. (1936*a*). *Physica*, **3**, 672; *Commun.*
  Phys. Lab. Univ. Leiden, no. 243*a*. [*78*]
BREMMER, H. and DE HAAS, W. J. (1936*b*). *Physica*, **3**, 692; *Commun.*
  Phys. Lab. Univ. Leiden, no. 243*c*. [*78*]
BRILLOUIN, L. (1933). *J. Phys. Radium*, **4**, 360. [*217*]
BRILLOUIN, L. (1935). *Proc. Roy. Soc.* A, **152**, 19. [*217*]

BROER, L. J. F. (1947). *Physica*, **13**, 473. *[194]*
BUCKEL, W. and HILSCH, R. (1950). *Z. Phys.* **128**, 324. *[38, 229]*
BUCKEL, W. and HILSCH, R. (1952). *Z. Phys.* **132**, 420. *[166]*
BURTON, E. F., TARR, F. G. A. and WILHELM, J. O. (1935). *Nature, Lond.*, **136**, 141. *[87]*
BURTON, E. F., WILHELM, J. O. and MISENER, A. D. (1934). *Trans. Roy. Soc. Can.* **28**, III, 65. *[166]*
CASIMIR, H. B. G. (1940). *Physica*, **7**, 887; *Commun. Phys. Lab. Univ. Leiden*, no. 261c. *[55, 139, 150]*
CASIMIR, H. B. G. and RADEMAKERS, A. (1947). *Physica*, **13**, 33; *Commun. Phys. Lab. Univ. Leiden*, no. 270d. *[87, 89, 91, 236]*
CHAMBERS, R. G. (1950). *Nature, Lond.*, **165**, 239. *[201]*
CLEMENT, J. R. and QUINNELL, E. H. (1950). *Phys. Rev.* **79**, 1028. *[62, 63]*
CLEMENT, J. R. and QUINNELL, E. H. (1952). *Phys. Rev.* **85**, 502. *[62]*
CLUSIUS, K. (1932). *Z. Elektrochem.* **38**, 312. *[4]*
CONDON, E. U. and MAXWELL, E. (1949). *Phys. Rev.* **76**, 578. *[51]*
COOK, D. B., ZEMANSKY, M. W. and BOORSE, H. A. (1950a). *Phys. Rev.* **79**, 1021. *[229]*
COOK, D. B., ZEMANSKY, M. W. and BOORSE, H. A. (1950b). *Phys. Rev.* **80**, 737. *[223, 227]*
DARBY, J., HATTON, J. and ROLLIN, B. V. (1950). *Proc. Phys. Soc.* A, **63**, 1181. *[229]*
DAUNT, J. G. (1937). *Phil. Mag.* **24**, 361. *[54]*
DAUNT, J. G. (1947). *Phys. Rev.* **72**, 89. *[197]*
DAUNT, J. G. (1950). *Phys. Rev.* **80**, 911. *[9]*
DAUNT, J. G. and HEER, C. V. (1949a). *Phys. Rev.* **76**, 715. *[223, 225]*
DAUNT, J. G. and HEER, C. V. (1949b). *Phys. Rev.* **76**, 1324. *[227]*
DAUNT, J. G., HORSEMAN, A. and MENDELSSOHN, K. (1939). *Phil. Mag.* **27**, 754. *[60, 225, 227]*
DAUNT, J. G., KEELEY, T. C. and MENDELSSOHN, K. (1937). *Phil. Mag.* **23**, 264. *[7]*
DAUNT, J. G. and MENDELSSOHN, K. (1937). *Proc. Roy. Soc.* A, **160**, 127. *[53, 59, 65, 71, 225, 227]*
DAUNT, J. G. and MENDELSSOHN, K. (1938). *Nature, Lond.*, **141**, 116. *[87]*
DAUNT, J. G. and MENDELSSOHN, K. (1946). *Proc. Roy. Soc.* A, **185**, 225. *[87]*
DAUNT, J. G., MILLER, A. R., PIPPARD, A. B. and SHOENBERG, D. (1948). *Phys. Rev.* **74**, 842. *[146, 197]*
DÉSIRANT, M. and SHOENBERG, D. (1948a). *Proc. Roy. Soc.* A, **194**, 63. *[36, 53, 98, 114, 115, 118, 130]*
DÉSIRANT, M. and SHOENBERG, D. (1948b). *Proc. Phys. Soc.* **60**, 413. *[53, 140, 147]*
VAN DIJK, H. and SHOENBERG, D. (1949). *Nature, Lond.*, **164**, 151. *[223]*
DOLECEK, R. L. and DE LAUNAY, J. (1949). *Phys. Rev.* **76**, 445. *[34]*
DRELL, S. D. (1951). *Phys. Rev.* **83**, 838. *[239]*
DUNAEV, J. A. (1947). *C.R. Acad. Sci. U.R.S.S.* **55**, 21. *[229]*
EHRENFEST, P. (1933). *Commun. Phys. Lab. Univ. Leiden*, Suppl. no. 75b. *[56, 60, 75]*
FABER, T. E. (1949). *Nature, Lond.*, **164**, 277. *[126]*

FABER, T. E. (1952). *Proc. Roy. Soc.* A (in the Press). [*125, 126, 128*]
FAIRBANK, W. M. (1949). *Phys. Rev.* **76**, 1106. [*11*]
FRASER, A. R. and SHOENBERG, D. (1949). *Proc. Camb. Phil. Soc.* **45**, 680. [*159, 189*]
FRITZ, J. J., GONZALEZ, O. D. and JOHNSTON, H. L. (1949). *Phys. Rev.* **76**, 580. [*51*]
FRITZ, J. J., GONZALEZ, O. D. and JOHNSTON, H. L. (1950). *Phys. Rev.* **80**, 894. [*51*]
FRÖHLICH, H. (1950). *Phys. Rev.* **79**, 845. [*4, 180, 185, 219*]
FRÖHLICH, H. (1951a). *Proc. Phys. Soc.* A, **64**, 129. [*219, 220*]
FRÖHLICH, H. (1951b). *Nature, Lond.*, **168**, 280. [*239*]
GALKIN, A. A. and BEZUGLYI, P. A. (1950). *J. Exp. Theor. Phys. U.S.S.R.* (Russian), **20**, 1145. [*128*]
GALKIN, A. A., KAN, YA. S. and LASAREW, B. G. (1950). *J. Exp. Theor. Phys. U.S.S.R.* **20**, 865. [*128*]
GALKIN, A. A. and LASAREW, B. G. (1948). *J. Exp. Theor. Phys. U.S.S.R.* (Russian), **18**, 833. [*128*]
GALKIN, A. A., LASAREW, B. G. and BEZUGLYI, P. A. (1950). *J. Exp. Theor. Phys. U.S.S.R.* (Russian), **20**, 987. [*128*]
GARFUNKEL, M. P. and SERIN, B. (1952). *Phys. Rev.* **85**, 834. [*237*]
GINSBURG, V. L. (1944). *J. Phys. U.S.S.R.* **8**, 148. [*159, 189*]
GINSBURG, V. L. (1945). *J. Phys. U.S.S.R.* **9**, 305. [*98, 172*]
GINSBURG, V. L. (1946). *Superconductivity* (in Russian), Moscow, Leningrad. [*180*]
GINSBURG, V. L. (1947). *J. Phys. U.S.S.R.* **11**, 93. [*161*]
GINSBURG, V. L. (1950). *Usp. Fiz. Nauk* (Russian), **12**, 169 and 332. [*86, 239*]
GINSBURG, V. L. (1951). *J. Exp. Theor. Phys. U.S.S.R.* (Russian), **21**, 979. [*239*]
GINSBURG, V. L. and LANDAU, L. D. (1950). *J. Exp. Theor. Phys. U.S.S.R.* (Russian), **20**, 1064. [*238*]
GOLIK, V. R., LASAREW, B. G. and CHOTKEWITSCH, W. I. (1949). *J. Exp. Theor. Phys.* (Russian), **19**, 202. [*229*]
GOODMAN, B. B. (1951a). *Nature, Lond.*, **167**, 111; *Low Temp. Symp. Bur. Stand. Wash.* [*3, 225, 226*]
GOODMAN, B. B. (1951b). *Oxford Conf. Low Temp.* p. 129. [*236*]
GOODMAN, B. B. and MENDOZA, E. B. (1951). *Phil. Mag.* **42**, 594. [*9, 121, 223, 225, 226*]
GOODMAN, B. B. and SHOENBERG, D. (1950). *Nature, Lond.*, **165**, 441. [*223, 225, 226, 227*]
GORTER, C. J. (1933). *Arch. Mus. Teyler*, **7**, 378. [*56*]
GORTER, C. J. (1935). *Physica*, **2**, 449. [*46*]
GORTER, C. J. (1949). *Physica*, **15**, 55; *Commun. Phys. Lab. Univ. Leiden*, Suppl. no. 98b. [*194*]
GORTER, C. J. and CASIMIR, H. B. G. (1934a). *Physica*, **1**, 306. [*56*]
GORTER, C. J. and CASIMIR, H. B. G. (1934b). *Phys. Z.* **35**, 963; *Z. Techn. Phys.* **15**, 539. [*179, 194*]
GRASSMANN, P. (1936). *Phys. Z.* **37**, 569. [*6*]
GREBENKEMPER, C. J. and HAGEN, J. P. (1951). *Low Temp. Symp. Bur. Stand. Wash.* [*11*]

GUTTMAN, L. and STOUT, J. W. (1951). *Low Temp. Symp. Bur. Stand. Wash.* [*228, 231*]

DE HAAS, W. J. (1933). *Leipziger Vorträge*, p. 59. [*122*]

DE HAAS, W. J., VAN AUBEL, E. and VOOGD, J. (1929*a*). *Commun. Phys. Lab. Univ. Leiden*, no. 197*a*. [*38, 228*]

DE HAAS, W. J., VAN AUBEL, E. and VOOGD, J. (1929*b*). *Commun. Phys. Lab. Univ. Leiden*, no. 197*b*. [*230, 231*]

DE HAAS, W. J., VAN AUBEL, E. and VOOGD, J. (1929*c*). *Commun. Phys. Lab. Univ. Leiden*, no. 197*c*. [*38*]

DE HAAS, W. J., VAN AUBEL, E. and VOOGD, J. (1929*d*). *Commun. Phys. Lab. Univ. Leiden*, no. 197*d*. [*230, 231*]

DE HAAS, W. J., VAN AUBEL, E. and VOOGD, J. (1930). *Commun. Phys. Lab. Univ. Leiden*, no. 208*a*. [*230, 231*]

DE HAAS, W. J. and DE BOER, J. (1932). *Commun. Phys. Lab. Univ. Leiden*, no. 220*a*. [*231*]

DE HAAS, W. J. and BREMMER, H. (1931). *Commun. Phys. Lab. Univ. Leiden*, no. 214*d*. [*78*]

DE HAAS, W. J. and BREMMER, H. (1932*a*). *Commun. Phys. Lab. Univ. Leiden*, no. 220*b*. [*78*]

DE HAAS, W. J. and BREMMER, H. (1932*b*). *Commun. Phys. Lab. Univ. Leiden*, no. 220*c*. [*78*]

DE HAAS, W. J. and BREMMER, H. (1936). *Physica*, 3, 687; *Commun. Phys. Lab. Univ. Leiden*, no. 243*b*. [*78*]

DE HAAS, W. J. and CASIMIR-JONKER, J. M. (1934). *Physica*, 1, 291; *Commun. Phys. Lab. Univ. Leiden*, no. 229*d*. [*53*]

DE HAAS, W. J. and CASIMIR-JONKER, J. M. (1935). *Commun. Phys. Lab. Univ. Leiden*, no. 233*c*. [*40*]

DE HAAS, W. J. and ENGELKES, A. D. (1937). *Physica*, 4, 325; *Commun. Phys. Lab. Univ. Leiden*, no. 247*d*. [*227*]

DE HAAS, W. J., ENGELKES, A. D. and GUINAU, O. A. (1937). *Physica*, 4, 595; *Commun. Phys. Lab. Univ. Leiden*, no. 247*e*. [*36*]

DE HAAS, W. J. and GUINAU, O. A. (1936). *Physica*, 3, 182 and 534; *Commun. Phys. Lab. Univ. Leiden*, nos. 241*a*, *b*. [*26, 53*]

DE HAAS, W. J. and JURRIAANSE, T. (1932). *Commun. Phys. Lab. Univ. Leiden*, no. 220*e*. [*38, 228*]

DE HAAS, W. J. and KINOSHITA, M. (1927). *Commun. Phys. Lab. Univ. Leiden*, no. 187*b*. [*7, 77*]

DE HAAS, W. J. and RADEMAKERS, A. (1940). *Physica*, 7, 992; *Commun. Phys. Lab. Univ. Leiden*, no. 261*e*. [*78, 81, 83, 85*]

DE HAAS, W. J., SIZOO, G. J. and ONNES, H. K. (1925). *Commun. Phys. Lab. Univ. Leiden*, no. 180*d*. [*121*]

DE HAAS, W. J. and VOOGD, J. (1928). *Commun. Phys. Lab. Univ. Leiden*, no. 191*d*. [*121*]

DE HAAS, W. J. and VOOGD, J. (1929*a*). *Commun. Phys. Lab. Univ. Leiden*, no. 199*c*. [*231*]

DE HAAS, W. J. and VOOGD, J. (1929*b*). *Commun. Phys. Lab. Univ. Leiden*, no. 199*d*. [*225*]

DE HAAS, W. J. and VOOGD, J. (1930). *Commun. Phys. Lab. Univ. Leiden*, no. 208*b*. [*39, 230, 231*]

248 SUPERCONDUCTIVITY

DE HAAS, W. J. and VOOGD, J. (1931a). *Commun. Phys. Lab. Univ. Leiden*, no. 212c. [8, 121]
DE HAAS, W. J. and VOOGD, J. (1931b). *Commun. Phys. Lab. Univ. Leiden*, no. 214b. [39, 228, 230]
DE HAAS, W. J. and VOOGD, J. (1931c). *Commun. Phys. Lab. Univ. Leiden*, no. 214c. [4, 5, 136, 212]
DE HAAS, W. J. and VOOGD, J. (1932). *Commun. Phys. Lab. Univ. Leiden*, Suppl. no. 73a. [232]
DE HAAS, W. J., VOOGD, J. and JONKER, J. M. (1934). *Physica*, **1**, 281; *Commun. Phys. Lab. Univ. Leiden*, no. 229c. [111, 112]
HALL, E. H. (1933). *Proc. Nat. Acad. Sci., Wash.*, **19**, 619. [49]
HANSEN, M. (1936). *Der Aufbau der Zweistofflegierungen.* Berlin: Springer. [232]
HATTON, J., ROLLIN, B. V. and SEYMOUR, E. F. W. (1951). *Proc. Phys. Soc.* A, **64**, 667. [229]
HEISENBERG, W. (1947). *Z. Naturf.* **2a**, 185. [180, 216]
HEISENBERG, W. (1948a). *Two Lectures.* Cambridge University Press. [180, 215, 216]
HEISENBERG, W. (1948b). *Z. Naturf.* **3a**, 65. [180, 215, 216]
HILSCH, R. (1951). *Oxford Conf. Low Temp.* p. 119. [239]
HIRSCHLAFF, E. (1937). *Proc. Camb. Phil. Soc.* **33**, 140. [7]
HOLM, R. and MEISSNER, W. (1932). *Z. Phys.* **74**, 715. [6]
HORN, F. H. and ZIEGLER, W. T. (1947). *J. Amer. Chem. Soc.* **69**, 2762. [228, 229]
HOUSTON, W. V. and SQUIRE, C. F. (1949a). *Science*, **109**, 439. [51]
HOUSTON, W. V. and SQUIRE, C. F. (1949b). *Phys. Rev.* **76**, 685. [51]
HUANG, K. (1951). *Proc. Phys. Soc.* A, **64**, 867. [239]
HUDSON, R. P. (1951). *Low Temp. Symp. Bur. Stand. Wash.*; *Proc. Phys. Soc.* A, **64**, 751. [229]
HULM, J. K. (1949). *Nature, Lond.*, **163**, 368. [78]
HULM, J. K. (1950). *Proc. Roy. Soc.* A, **204**, 98. [78, 79, 83, 215, 216]
HULM, J. K. and MATTHIAS, B. T. (1951a). *Phys. Rev.* **82**, 273. [228, 229]
HULM, J. K. and MATTHIAS, B. T. (1951b). *Low Temp. Symp. Bur. Stand. Wash.* [228]
JACKSON, L. C. and PRESTON-THOMAS, H. (1950). *Phil. Mag.* **41**, 1284. [47, 223, 225]
JURRIAANSE, T. (1935). *Z. Kristallogr.* **90**, 322. [38, 228]
JUSTI, E. (1946). *Neue Phys. Blätter*, **8**, 207. [225]
JUSTI, E. and ZICKNER, G. (1941). *Phys. Z.* **42**, 258. [49]
KAN, L. S., SUDOVSTOV, A. I. and LASAREW, B. G. (1948). *J. Exp. Theor. Phys. U.S.S.R.* (Russian), **18**, 825. [75, 76, 77]
KAN, L. S., SUDOVSTOV, A. I. and LASAREW, B. G. (1949). *Doklady Akad. Nauk S.S.S.R.* (Russian), **69**, 173. [75, 76]
KEELEY, T. C. and MENDELSSOHN, K. (1936). *Proc. Roy. Soc.* A, **154**, 378. [53]
KEESOM, W. H. (1924). *Rapp et Disc. 4e Congr. Phys. Solvay*, p. 288. [56, 59]
KEESOM, W. H. (1933). *Commun. Phys. Lab. Univ. Leiden*, no. 224c. [225, 227]

KEESOM, W. H. (1934). *Physica*, 1, 123; but see also postscript of *Commun. Phys. Lab. Univ. Leiden*, no. 230a. [*225*]

KEESOM, W. H. (1935). *Physica*, 2, 35; *Commun. Phys. Lab. Univ. Leiden*, no. 234f. [*41*]

KEESOM, W. H. and DÉSIRANT, M. (1941). *Physica*, 8, 273; *Commun. Phys. Lab. Univ. Leiden*, no. 257b. [*62*]

KEESOM, W. H. and KOK, J. A. (1932). *Commun. Phys. Lab. Univ. Leiden*, no. 221e. [*60, 61, 62*]

KEESOM, W. H. and KOK, J. A. (1934a). *Physica*, 1, 175; *Commun. Phys. Lab. Univ. Leiden*, no. 230c. [*62, 63*]

KEESOM, W. H. and KOK, J. A. (1934b). *Physica*, 1, 503; *Commun. Phys. Lab. Univ. Leiden*, no. 230e. [*60, 67*]

KEESOM, W. H. and KOK, J. A. (1934c). *Physica*, 1, 595; *Commun. Phys. Lab. Univ. Leiden*, no. 232a. [*60, 67, 68, 69, 70*]

KEESOM, W. H. and VAN LAER, P. H. (1937). *Physica*, 4, 487; *Commun. Phys. Lab. Univ. Leiden*, no. 248c. [*60*]

KEESOM, W. H. and VAN LAER, P. H. (1938). *Physica*, 5, 193; *Commun. Phys. Lab. Univ. Leiden*, no. 252b. [*60, 61, 62, 63, 69, 70*]

KEESOM, W. H. and MATTHIJS, C. J. (1938a). *Physica*, 5, 1; *Commun. Phys. Lab. Univ. Leiden*, no. 250d. [*86, 89, 90*]

KEESOM, W. H. and MATTHIJS, C. J. (1938b). *Physica*, 5, 437; *Commun. Phys. Lab. Univ. Leiden*, no. 252e. [*89, 91, 92*]

KEESOM, W. H. and ONNES, H. K. (1924). *Commun. Phys. Lab. Univ. Leiden*, no. 174b. [*6*]

KHAIKIN, M. S. (1950). *Doklady Akad. Nauk S.S.S.R.* (Russian), 75, 661. [*11*]

KIKOIN, I. K. and GOOBAR, S. V. (1938). *C.R. Acad. Sci. U.R.S.S.* 19, 249. [*50*]

KIKOIN, I. K. and GOOBAR, S. V. (1940). *J. Phys. U.S.S.R.* 3, 333. [*50*]

KLEIN, O. (1952). *Nature, Lond.*, 169, 578. [*239*]

KOHN, W. and VACHASPATI (1951). *Phys. Rev.* 83, 462. [*239*]

KOK, J. A. (1934). *Physica*, 1, 1103; *Commun. Phys. Lab. Univ. Leiden*, Suppl. no. 77a. [*63*]

KOK, J. A. and KEESOM, W. H. (1937). *Physica*, 4, 835; *Commun. Phys. Lab. Univ. Leiden*, no. 248e. [*62*]

KOPPE, H. (1947). *Ann. Phys., Lpz.*, (6) 1, 405. [*236*]

KOPPE, H. (1948). *Z. Naturf.* 3a, 1. [*217*]

KOPPE, H. (1950). *Ergebn. Exakt. Naturwiss.* 23, 283. [*162, 216, 218*]

KUPER, C. G. (1951). *Phil. Mag.* 42, 961. [*99, 103, 114, 115, 237*]

KÜRTI, N. and SIMON, F. E. (1935). *Proc. Roy. Soc.* A, 151, 610. [*3, 55, 225*]

LANDAU, L. D. (1930). *Z. Phys.* 64, 629. [*185*]

LANDAU, L. D. (1937). *Phys. Z. Sowjet.* 11, 129. [*98, 237*]

LANDAU, L. D. (1938). *Nature, Lond.*, 141, 688. [*98, 100*]

LANDAU, L. D. (1943). *J. Phys. U.S.S.R.* 7, 99. [*98, 100, 101, 102, 115, 119*]

LANDAU, L. D. and LIFSHITZ, E. M. (1938). *Statistical Physics.* Oxford: Clarendon Press. [*60*]

LASAREW, B. G. and GALKIN, A. A. (1944). *J. Phys. U.S.S.R.* 8, 371. [*39, 125*]

LASAREW, B. G., GALKIN, A. A. and CHOTKEWITSCH, W. I. (1947). *Doklady Akad. Nauk S.S.S.R.* (Russian), **55**, 817. [*128*]

LASAREW, B. G. and KAN, L. S. (1944a). *J. Phys. U.S.S.R.* **8**, 193. [*75*]

LASAREW, B. G. and KAN, L. S. (1944b). *J. Phys. U.S.S.R.* **8**, 361. [*75*]

LASAREW, B. G. and NAKHUTIN, I. E. (1942). *J. Phys. U.S.S.R.* **6**, 116. [*38, 231*]

LASAREW, B. G. and SUDOVSTOV, A. I. (1949). *Doklady Akad. Nauk S.S.S.R.* (Russian), **69**, 345. [*76, 77*]

VON LAUE, M. (1932). *Phys. Z.* **33**, 793. [*49, 111*]

VON LAUE, M. (1938). *Ann. Phys., Lpz.,* **32**, 71. [*214*]

VON LAUE, M. (1942). *Phys. Z.* **43**, 274. [*214*]

VON LAUE, M. (1947). *Naturwissenschaften,* **34,** 186. [*159*]

VON LAUE, M. (1948). *Ann. Phys., Lpz.,* **3**, 31. [*159, 189*]

VON LAUE, M. (1949). *Theorie der Supraleitung,* 2nd ed. Berlin-Göttingen-Heidelberg: Springer. [*24, 180, 181*]

DE LAUNAY, J. (1949). *Naval Res. Lab. Tech. Rep.* P-3441. [*34*]

LAURMANN, E. and SHOENBERG, D. (1947). *Nature, Lond.,* **160**, 747. [*55, 139, 151, 159*]

LAURMANN, E. and SHOENBERG, D. (1949). *Proc. Roy. Soc.* A, **198**, 560. [*55, 139, 150, 151, 153, 154, 155, 156, 159, 162, 190, 225*]

LIFSHITZ, E. M. and SHARVIN, YU. V. (1951). *Doklady Akad. Nauk S.S.S.R.* (Russian), **79**, 783. [*236*]

LIFSHITZ, I. M. (1950). *J. Exp. Theor. Phys. U.S.S.R.* (Russian), **20**, 834. [*128*]

LOCK, J. M. (1950). Ph.D. Thesis, Cambridge Univ. [*209, 211*]

LOCK, J. M. (1951). *Proc. Roy. Soc.* A, **208**, 391. [*140, 148, 150, 162, 163, 165, 166, 168, 170*]

LOCK, J. M., PIPPARD, A. B. and SHOENBERG, D. (1951). *Proc. Camb. Phil. Soc.* **47**, 811. [*12, 223, 225, 226*]

LONDON, F. (1935). *Proc. Roy. Soc.* A, **152**, 24. [*184*]

LONDON, F. (1936). *Physica,* **3**, 450. [*25*]

LONDON, F. (1937). *Une conception nouvelle de la supraconductibilité.* Paris: Hermann et Cie. [*131, 135*]

LONDON, F. (1948). *Phys. Rev.* **74**, 562. [*184*]

LONDON, F. (1950). *Superfluids,* **1**. New York: Wiley. [*180, 184*]

LONDON, F. and LONDON, H. (1935a). *Proc. Roy. Soc.* A, **149**, 71. [*138, 179, 180*]

LONDON, F. and LONDON, H. (1935b). *Physica,* **2**, 341. [*180*]

LONDON, H. (1934). *Nature, Lond.,* **133**, 497. [*197*]

LONDON, H. (1935). *Proc. Roy. Soc.* A, **152**, 650. [*46, 122, 175*]

LONDON, H. (1936). *Proc. Roy. Soc.* A, **155**, 102. [*187*]

LONDON, H. (1940). *Proc. Roy. Soc.* A, **176**, 522. [*11, 197, 200*]

MACDONALD, D. K. C. and MENDELSSOHN, K. (1949). *Proc. Roy. Soc.* A, **200**, 66. [*9*]

MCLENNAN, J. C., ALLEN, J. F. and WILHELM, J. O. (1930a). *Trans. Roy. Soc. Can.* **24**, III, 25. [*230, 231*]

MCLENNAN, J. C., ALLEN, J. F. and WILHELM, J. O. (1930b). *Trans. Roy. Soc. Can.* **24**, III, 53. [*229, 230, 231*]

MCLENNAN, J. C., ALLEN, J. F. and WILHELM, J. O. (1931a). *Trans. Roy. Soc. Can.* **25**, III, 1. [*7, 77*]

McLennan, J. C., Allen, J. F. and Wilhelm, J. O. (1931 b). Trans. Roy. Soc. Can. 25, III, 13. [228]
McLennan, J. C., Allen, J. F. and Wilhelm, J. O. (1932). Phil. Mag. 13, 1196. [230, 231]
McLennan, J. C., Howlett, L. E. and Wilhelm, J. O. (1929). Trans. Roy. Soc. Can. 23, III, 287. [230]
McLennan, J. C., Hunter, R. G. and McLeod, J. H. (1930). Trans. Roy. Soc. Can. 24, III, 3. [7]
McLennan, J. C., McLeod, J. H. and Wilhelm, J. O. (1929). Trans. Roy. Soc. Can. 23, III, 269. [7]
McLennan, J. C., Niven, C. D. and Wilhelm, J. O. (1928). Phil. Mag. 6, 678. [230, 231]
Makinson, R. E. B. (1938). Proc. Camb. Phil. Soc. 34, 474. [80]
Maxwell, E. (1950a). Phys. Rev. 78, 477. [12]
Maxwell, E. (1950b). Phys. Rev. 79, 173. [12]
Maxwell, E. (1951). Low Temp. Symp. Bur. Stand. Wash. [236]
Maxwell, E., Marcus, P. M. and Slater, J. C. (1949). Phys. Rev. 76, 1332. [11, 201]
Maxwell, J. C. (1892). Electricity and Magnetism, 3rd ed. Oxford. [22]
Meissner, W. (1927). Z. ges. Kältenindustr. 34, 197. [86]
Meissner, W. (1928). Phys. Z. 29, 897. [225]
Meissner, W. (1929a). Z. Phys. 58, 570. [38, 229]
Meissner, W. (1929b). Naturwissenschaften, 17, 390. [225]
Meissner, W. (1930a). Z. Phys. 60, 181. [223, 225]
Meissner, W. (1930b). Z. Phys. 61, 191. [223]
Meissner, W. and Franz, H. (1930a). Z. Phys. 63, 558. [225]
Meissner, W. and Franz, H. (1930b). Naturwissenschaften, 18, 418; Z. Phys. 65, 30. [228, 229]
Meissner, W., Franz, H. and Westerhoff, H. (1932a). Z. Phys. 75, 521. [228, 229]
Meissner, W., Franz, H. and Westerhoff, H. (1932b). Ann. Phys., Lpz., 13, 505. [228, 230, 231]
Meissner, W., Franz, H. and Westerhoff, H. (1932c). Ann. Phys., Lpz., 13, 555. [223, 225]
Meissner, W., Franz, H. and Westerhoff, H. (1932d). Ann. Phys., Lpz., 13, 967. [230]
Meissner, W., Franz, H. and Westerhoff, H. (1933). Ann. Phys., Lpz., 17, 593. [228, 230, 231]
Meissner, W. and Heidenreich, F. (1936). Phys. Z., Lpz., 37, 449. [49]
Meissner, W. and Ochsenfeld, R. (1933). Naturwissenschaften, 21, 787. [13, 16, 52]
Meissner, W. and Steiner, K. (1932). Z. Phys. 76, 201. [7]
Meissner, W. and Voigt, B. (1930a). Ann. Phys., Lpz., 7, 761. [225]
Meissner, W. and Voigt, B. (1930b). Ann. Phys., Lpz., 7, 892. [225]
Meissner, W. and Westerhoff, H. (1934). Z. Phys. 87, 206. [223, 225]
Mendelssohn, K. (1935). Proc. Roy. Soc. A, 152, 34. [44, 72]
Mendelssohn, K. (1936). Proc. Roy. Soc. A, 155, 558. [53]
Mendelssohn, K. (1941). Nature, Lond., 148, 316. [62]
Mendelssohn, K. (1952). Nature, Lond., 169, 266. [71]

MENDELSSOHN, K. and BABBITT, J. D. (1935). *Proc. Roy. Soc.* A, **151**, 316.
[*26, 36, 53*]
MENDELSSOHN, K. and DAUNT, J. G. (1937). *Nature, Lond.*, **139**, 473.
[*225*]
MENDELSSOHN, K. and MOORE, J. R. (1934). *Nature, Lond.*, **133**, 413. [*70*]
MENDELSSOHN, K. and MOORE, J. R. (1935a). *Nature, Lond.*, **135**, 826.
[*40, 53*]
MENDELSSOHN, K. and MOORE, J. R. (1935b). *Proc. Roy. Soc.* A, **151**,
334. [*72*]
MENDELSSOHN, K. and OLSEN, J. L. (1950a). *Proc. Phys. Soc.* A, **63**, 2.
[*78, 81, 83, 85*]
MENDELSSOHN, K. and OLSEN, J. L. (1950b). *Proc. Phys. Soc.* A, **63**, 1182.
[*78, 84*]
MENDELSSOHN, K. and OLSEN, J. L. (1950c). *Phys. Rev.* **80**, 859. [*78, 85*]
MENDELSSOHN, K. and PONTIUS, R. B. (1936a). *Physica*, **3**, 327. [*36, 53,
121*]
MENDELSSOHN, K. and PONTIUS, R. B. (1936b). *Nature, Lond.*, **138**, 29.
[*36, 53*]
MENDELSSOHN, K. and PONTIUS, R. B. (1937). *Phil. Mag.* **24**, 777. [*78*]
MENDOZA, E. and THOMAS, J. G. (1951). *Phil. Mag.* **42**, 291. [*225*]
MESHKOVSKY, A. G. (1949). *J. Exp. Theor. Phys. U.S.S.R.* (Russian),
19, 1. [*103, 104, 105, 108*]
MESHKOVSKY, A. G. and SHALNIKOV, A. I. (1947a). *J. Phys. U.S.S.R.*
11, 1. [*53, 103, 104, 106, 110*]
MESHKOVSKY, A. G. and SHALNIKOV, A. I. (1947b). *J. Exp. Theor. Phys.
U.S.S.R.* (Russian), **17**, 851. [*53, 103, 104, 105, 109, 110*]
MISENER, A. D. (1936). *Canad. J. Res.* **14**, 25. [*166, 177*]
MISENER, A. D. (1938). *Proc. Roy. Soc.* A, **166**, 43. [*114, 117*]
MISENER, A. D. (1939). *Proc. Camb. Phil. Soc.* **35**, 95. [*112*]
MISENER, A. D. (1940). *Proc. Roy. Soc.* A, **174**, 262. [*60, 225, 226*]
MISENER, A. D., SMITH, H. GRAYSON, and WILHELM, J. O. (1935). *Trans.
Roy. Soc. Can.* **29**, III, 13. [*166*]
MISENER, A. D. and WILHELM, J. O. (1935). *Trans. Roy. Soc. Can.* **29**, III, 1.
[*166*]
MOTT, N. F. and JONES, H. (1936). *The Theory and Properties of Metals
and Alloys.* Oxford. [*63*]
NAKHUTIN, I. E. (1938). *J. Exp. Theor. Phys. U.S.S.R.* (Russian), **8**, 713.
[*112*]
OLSEN-BÄR, M. (1951). *Oxford Conf. Low Temp.* p. 118. [*236*]
ONNES, H. K. (1911a). *Commun. Phys. Lab. Univ. Leiden*, no. 119b. [*1*]
ONNES, H. K. (1911b). *Commun. Phys. Lab. Univ. Leiden*, no. 120b.
[*1, 225*]
ONNES, H. K. (1911c). *Commun. Phys. Lab. Univ. Leiden*, no. 122b.
[*1, 225*]
ONNES, H. K. (1911d). *Commun. Phys. Lab. Univ. Leiden*, no. 124c. [*3*]
ONNES, H. K. (1913a). *Commun. Phys. Lab. Univ. Leiden*, Suppl. no. 34.
[*1, 2, 10*]
ONNES, H. K. (1913b). *Commun. Phys. Lab. Univ. Leiden*, nos. 133a,b,c. [*5*]
ONNES, H. K. (1913c). *Commun. Phys. Lab. Univ. Leiden*, no. 133d. [*10,
225, 231*]

ONNES, H. K. (1914a). *Commun. Phys. Lab. Univ. Leiden*, no. 139f. [8]

ONNES, H. K. (1914b). *Commun. Phys. Lab. Univ. Leiden*, nos. 140b, c. 141b. [6]

ONNES, H. K. and HOF, K. (1914). *Commun. Phys. Lab. Univ. Leiden*, no. 142b. [49]

ONNES, H. K. and TUYN, W. (1922a). *Commun. Phys. Lab. Univ. Leiden*, no. 160a. [225]

ONNES, H. K. and TUYN, W. (1922b). *Commun. Phys. Lab. Univ. Leiden*, no. 160b. [12]

PARKINSON, D. H., SIMON, F. E. and SPEDDING, F. H. (1951). *Proc. Roy. Soc.* A, **207**, 137. [62, 63, 225]

PEIERLS, R. (1936). *Proc. Roy. Soc.* A, **155**, 613. [25, 66]

PIMENTEL, G. C. and SHELINE, R. K. (1949). *J. Chem. Phys.* **17**, 644. [90]

PIPPARD, A. B. (1947a). *Proc. Roy. Soc.* A, **191**, 370. [11, 159, 201]

PIPPARD, A. B. (1947b). *Proc. Roy. Soc.* A, **191**, 385. [11, 157, 198, 201]

PIPPARD, A. B. (1947c). *Proc. Roy. Soc.* A, **191**, 399. [11, 139, 150, 156, 157, 198, 202]

PIPPARD, A. B. (1950a). *Proc. Roy. Soc.* A, **203**, 98. [11, 156, 158, 159, 160, 191, 198, 201, 205]

PIPPARD, A. B. (1950b). *Proc. Roy. Soc.* A, **203**, 195. [11, 158, 198, 203, 206]

PIPPARD, A. B. (1950c). *Proc. Roy. Soc.* A, **203**, 210. [4, 46, 130, 161, 207, 208, 212, 237]

PIPPARD, A. B. (1950d). *Phil. Mag.* **41**, 243. [71, 128, 175]

PIPPARD, A. B. (1951a). *Proc. Camb. Phil. Soc.* **47**, 617. [46, 98, 171, 176, 207, 213]

PIPPARD, A. B. (1951b). *Oxford Conf. Low Temp.* p. 118. [209]

PIPPARD, A. B. (1952). *Phil. Mag.* **43**, 273. [147, 165, 210]

PIPPARD, A. B. and PULLAN, G. T. (1952). *Proc. Camb. Phil. Soc.* **48**, 188. [236]

PONTIUS, R. B. (1937). *Phil. Mag.* **24**, 787. [169]

PULLAN, G. T. (1951). *Oxford Conf. Low Temp.* p. 133. [236]

RADEMAKERS, A. (1949). *Physica*, **15**, 849, *Commun. Phys. Lab. Univ. Leiden*, no. 279b. [78, 82]

RAMANATHAN, K. G. (1952). *Proc. Phys. Soc.* A (in the Press). [7]

REUTER, G. E. H. and SONDHEIMER, E. H. (1948). *Proc. Roy. Soc.* A, **195**, 336. [198, 201]

REYNOLDS, C. A. and SERIN, B. (1951). *Oxford Conf. Low Temp.* p. 116. [236]

REYNOLDS, C. A., SERIN, B. and NESBITT, L. B. (1951). *Phys. Rev.* **84**, 691. [12]

REYNOLDS, C. A., SERIN, B., WRIGHT, W. H. and NESBITT, L. B. (1950). *Phys. Rev.* **78**, 487. [12]

REYNOLDS, J. M. and LANE, C. T. (1950). *Phys. Rev.* **79**, 405. [228]

RJABININ, J. N. and SCHUBNIKOW, L. W. (1934). *Phys. Z. Sowjet.* **5**, 641; **6**, 557. [53]

RJABININ, J. N. and SCHUBNIKOW, L. W. (1935). *Nature, Lond.*, **135**, 581; *Phys. Z. Sowjet.* **7**, 122. [40, 41, 43, 230]

RUTGERS, A. J. (1934). *Physica*, **1**, 1055. [56]

RUTGERS, A. J. (1936). *Physica*, **3**, 999. [*56*]

SCHAFROTH, M. R. (1951). *Helv. Phys. Acta*, **24**, 645. [*239*]

SCHUBNIKOW, L. W. and ALEKSEYEVSKY, N. E. (1936). *Nature, Lond.*, **138**, 804. [*133*]

SCHUBNIKOW, L. W. and CHOTKEWITSCH, W. I. (1934). *Phys. Z. Sowjet.* **6**, 605. [*72*]

SCHUBNIKOW, L. W. and CHOTKEWITSCH, W. I. (1936). *Phys. Z. Sowjet.* **10**, 231. [*29, 31, 34*]

SCHUBNIKOW, L. W., CHOTKEWITSCH, W. I., SCHEPELEW, J. D. and RJABININ, J. N. (1936). *Phys. Z. Sowjet.* **10**, 165. [*41, 44, 230, 231*]

SCHUBNIKOW, L. W. and NAKHUTIN, I. E. (1937). *Nature, Lond.*, **139**, 589. [*112*]

SCOTT, R. B. (1948). *Bur. Stand. J. Res., Wash.*, **41**, 581. [*10, 134, 227*]

SEITZ, F. (1940). *Modern Theory of Solids.* New York: McGraw Hill. [*90*]

SERIN, B. and REYNOLDS, C. A. (1952). *Phys. Rev.* **85**, 938. [*128*]

SERIN, B., REYNOLDS, C. A., FELDMEIER, J. R. and GARFUNKEL, M. P. (1951). *Phys. Rev.* **84**, 802. [*128*]

SHALNIKOV, A. I. (1938). *Nature, Lond.*, **142**, 74. [*138, 166, 177*]

SHALNIKOV, A. I. (1940a). *J. Exp. Theor. Phys. U.S.S.R.* (Russian), **10**, 630. [*138, 166, 177, 178*]

SHALNIKOV, A. I. (1940b). *J. Phys. U.S.S.R.* **2**, 477. [*36*]

SHALNIKOV, A. I. (1942). *J. Phys. U.S.S.R.* **6**, 53. [*36*]

SHALNIKOV, A. I. (1945). *J. Phys. U.S.S.R.* **9**, 202. [*53, 103*]

SHALNIKOV, A. I. and SHARVIN, YU. V. (1948). *J. Exp. Theor. Phys. U.S.S.R.* (Russian), **18**, 102; *Bull. Acad. Sci. U.S.S.R.*, Phys. Ser. (Russian), **12**, 195. [*155*]

SHARVIN, YU. V. (1945). *J. Phys. U.S.S.R.* **9**, 350. [*4*]

SHARVIN, YU. V. (1951). *J. Exp. Theor. Phys. U.S.S.R.* (Russian), **21**, 658. [*237*]

SHOENBERG, D. (1935). *Proc. Roy. Soc.* A, **152**, 10. [*27, 54*]

SHOENBERG, D. (1936). *Proc. Roy. Soc.* A, **155**, 712. [*26, 27, 29, 31, 54*]

SHOENBERG, D. (1937a). *Proc. Camb. Phil. Soc.* **33**, 260. [*34, 35, 44, 54, 121*]

SHOENBERG, D. (1937b). *Proc. Camb. Phil. Soc.* **33**, 559. [*54, 112*]

SHOENBERG, D. (1937c). *Proc. Camb. Phil. Soc.* **33**, 577. [*30, 31, 226*]

SHOENBERG, D. (1938). *Nature, Lond.*, **142**, 874. [*38, 227, 228, 229*]

SHOENBERG, D. (1939). *Nature, Lond.*, **143**, 434. [*143*]

SHOENBERG, D. (1940a). *Proc. Roy. Soc.* A, **175**, 49. [*140, 143, 144, 145, 146, 162, 164, 170, 171*]

SHOENBERG, D. (1940b). *Proc. Camb. Phil. Soc.* **36**, 84. [*9, 121, 223, 225*]

SHOENBERG, D. (1947). *Phys. Soc. Camb. Conference Rep.* **2**, 85. [*53*]

SILSBEE, F. B. (1916). *J. Wash. Acad. Sci.* **6**, 597. [*10*]

SIMON, I. (1949). *R.L.E. Tech. Rep.* no. 126, M.I.T. [*11*]

SIMON, I. (1950). *Phys. Rev.* **77**, 384. [*11*]

SIXTUS, K. J. and TONKS, L. (1931). *Phys. Rev.* **37**, 930. [*127*]

SIZOO, G. J. (1926). Dissertation, Leiden. [*49*]

SIZOO, G. J., DE HAAS, W. J. and ONNES, H. K. (1925). *Commun. Phys. Lab. Univ. Leiden*, no. 180c. [*75, 121*]

SIZOO, G. J. and ONNES, H. K. (1925). *Commun. Phys. Lab. Univ. Leiden*, no. 180b. [*75*]

SMITH, H. GRAYSON and WILHELM, J. O. (1936). *Proc. Roy. Soc.* A, **157**, 132. [*29*]

SQUIRE, C. F. and LOVE, W. F. (1949). *Int. Conf. on Phys. of Very Low Temps.*, M.I.T., p. 102. [*50*]

STARK, J. and STEINER, K. (1937). *Phys. Z.* **38**, 277. [*19*]

STEELE, M. C. (1951). *Phys. Rev.* **81**, 262. [*89, 90, 91, 93, 236*]

STEINER, K. and GRASSMANN, P. (1935). *Phys. Z.* **36**, 527. [*87*]

STONER, E. C. (1945). *Phil. Mag.* **36**, 803. [*22*]

STOUT, J. W. and GUTTMAN, L. (1950). *Phys. Rev.* **79**, 396. [*38, 46, 231*]

TARR, F. G. A. and WILHELM, J. O. (1934). *Trans. Roy. Soc. Can.* **28**, III, 61. [*52*]

TISZA, L. (1950). *Phys. Rev.* **80**, 717. [*219*]

TISZA, L. (1951). *Phys. Rev.* **84**, 163. [*219*]

TUYN, W. and ONNES, H. K. (1923). *Commun. Phys. Lab. Univ. Leiden,* no. 167*a*. [*225*]

TUYN, W. and ONNES, H. K. (1926). *Commun. Phys. Lab. Univ. Leiden,* no. 174*a*. [*8, 10*]

WEBBER, R. T. (1947). *Phys. Rev.* **72**, 1241. [*47, 227*]

WEBBER, R. T., REYNOLDS, J. M. and McGUIRE, T. R. (1949). *Phys. Rev.* **76**, 293. [*227*]

WEBBER, R. T. and SPOHR, D. A. (1951). *Phys. Rev.* **84**, 384. [*236*]

WEBBER, R. T. and STEELE, M. C. (1950). *Phys. Rev.* **79**, 1028. [*91, 92, 236*]

WENTZEL, G. (1951). *Phys. Rev.* **83**, 168. [*239*]

WESTERFIELD, E. C. (1939). *Phys. Rev.* **55**, 319. [*77*]

WEXLER, A. and CORAK, W. S. (1949). *Phys. Rev.* **76**, 432. [*51*]

WEXLER, A. and CORAK, W. S. (1950*a*). *Phys. Rev.* **78**, 260. [*51*]

WEXLER, A. and CORAK, W. S. (1950*b*). *Phys. Rev.* **79**, 737. [*223, 225*]

ZAVARITSKI, N. V. (1951). *Doklady Akad. Nauk S.S.S.R.* (Russian), **78**, 665. [*237*]

ZIEGLER, W. T. and YOUNG, R. A. (1951). *Oxford Conf. Low Temp.* p. 124. [*228*]

SUBJECT INDEX

258 SUBJECT INDEX